Advances in
CHROMATOGRAPHY

Volume **52**

Advances in
CHROMATOGRAPHY

EDITORS
Eli Grushka • Nelu Grinberg

CRC Press
Taylor & Francis Group
Boca Raton London New York

CRC Press is an imprint of the
Taylor & Francis Group, an **informa** business

CRC Press
Taylor & Francis Group
6000 Broken Sound Parkway NW, Suite 300
Boca Raton, FL 33487-2742

First issued in paperback 2019

© 2014 by Taylor & Francis Group, LLC
CRC Press is an imprint of Taylor & Francis Group, an Informa business

No claim to original U.S. Government works

ISBN-13: 978-1-4822-2350-7 (hbk)
ISBN-13: 978-0-367-37877-6 (pbk)

Visit the Taylor & Francis Web site at
http://www.taylorandfrancis.com

and the CRC Press Web site at
http://www.crcpress.com

Table of Contents

Contributors

Ryan D. Cohen
Analytical Chemistry Department
Merck Research Laboratories
Rahway, New Jersey, USA

Jennifer L. Colangelo
Drug Safety
Pfizer Worldwide Research
    and Development
Groton, Connecticut, USA

Markus Juza
Institute of Organic Chemistry
University of Tübingen
Tübingen, Germany

Lei Ling
School of Chemical Engineering
Purdue University
West Lafayette, Indiana, USA

Yong Liu
Analytical Chemistry Department
Merck Research Laboratories
Rahway, New Jersey, USA

P. Nikitas
Laboratory of Physical Chemistry
Department of Chemistry
Aristotle University of Thessaloniki
Thessaloniki, Greece

A. Pappa-Louisi
Laboratory of Physical Chemistry
Department of Chemistry
Aristotle University of Thessaloniki
Thessaloniki, Greece

Volker Schurig
Institute of Organic Chemistry
University of Tübingen
Tübingen, Germany

Nien-Hwa Linda Wang
School of Chemical Engineering
Purdue University
West Lafayette, Indiana, USA

Ch. Zisi
Laboratory of Physical Chemistry
Department of Chemistry
Aristotle University of Thessaloniki
Thessaloniki, Greece

Advances in
CHROMATOGRAPHY

1 Advances in Aerosol-Based Detectors

Ryan D. Cohen and Yong Liu

1.1 INTRODUCTION

When analytes lack an ultraviolet (UV) chromophore, alternative detection methods are necessary for high-performance liquid chromatography (HPLC), capillary electrophoresis (CE), and supercritical fluid chromatography (SFC). Common alternatives to UV detection include refractive index (RI), mass spectrometry (MS), and aerosol-based detection. Both RI and MS have a few major drawbacks where aerosol-based detection is more appropriate. RI detectors are known to be generally unreliable, are incompatible with gradient elution, and suffer from poor sensitivity [1]. MS detectors are primarily limited by their ability to ionize the analyte, are expensive, often require a skilled operator, are frequently susceptible to matrix

1

effects, and exhibit a compound-specific, nonuniform response, although both the expense and ease of use have been dropping rapidly in recent years [2,3]. There are also several less-commonly employed UV alternatives, including chemilumines-cence (CL) nitrogen detection, nuclear magnetic resonance, infrared, indirect UV, conductivity, and amperometric detection. All these suffer from various drawbacks to making them a preferred alternative to UV, but they have advantages for particu-lar applications.

Some attractive qualities of aerosol-based detectors include ruggedness, relatively low cost, multiple commercially available models, low detection limits, acceptable precision, and a response independent of molecular structure. There are a few well-known limitations of this technology, such as nonlinear response, requirement for volatile mobile phases, and negligible-to-poor sensitivity for volatile and some semi-volatile molecules. In the following sections, these limitations are more thoroughly discussed along with recent innovations attempting to overcome them. In addition, several important problems where aerosol-based detectors have been found to be particularly useful are discussed.

Since 2000, several reviews and book chapters have been published on aerosol-based detectors, but none have covered all three major types [4–10]. In this chapter, we attempt a comprehensive review of the history, development, and theory behind this important class of detectors.

1.2 BACKGROUND AND HISTORY

The first aerosol-based detector to be described in the literature for liquid chroma-tography (LC) was the evaporative light-scattering detector (ELSD) in 1966 by Ford and Kennard [11]. However, the ELSD was not further discussed until a publication in 1978 by Charlesworth [12]. The first commercially available detector was devel-oped by Applied Chromatography System Limited in the early 1980s [13].

The need to improve on insufficient sensitivity of ELSDs led Allen and Koropchak to develop the condensation nucleation light-scattering detector (CNLSD) in the early to mid-1990s [14,15]. It took over a decade for this technology to be commercialized by Quant Technologies in 2009. To increase marketability, the name was changed to the nanoquantity analyte detector (NQAD) to highlight detection limits attainable at the nanogram-per-milliliter level.

In 2002, the charged aerosol detector (CAD) was first reported in a seminal paper by Dixon and Peterson; they referred to their method as aerosol charge detection [16]. They combined the operations of nebulizing and evaporating a separation's effluent with detection via an electrical aerosol analyzer, an established technology that had been used for aerosol sizing since the 1970s [17]. Subsequently, a modified version of the instrument they described was brought to market by ESA in 2004 as the Corona CAD™ [18,19].

A fourth type, the chemiluminescence aerosol detector (herein referred to as the CLAD), was first described in the literature in 2005 by Zhang and coworkers [20,21], but it has yet to be commercialized or further developed.

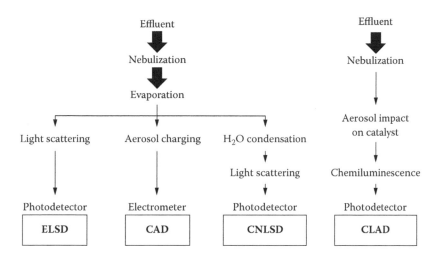

FIGURE 1.1 Diagram of key components and processes illustrating differences between aerosol-based detectors. ELSD, evaporative light-scattering detector; CAD, charged aerosol detector; CNLSD, condensation nucleation light-scattering detector; and CLAD, chemiluminescence aerosol detector.

1.3 TYPES OF AEROSOL-BASED DETECTORS

A diagram listing the key components of the four types of aerosol-based detectors is shown in Figure 1.1. The basis of these detectors is the generation of an aerosol via nebulization. Most frequently, this is accomplished via pneumatic nebulization, which uses a gas stream, often purified N_2, to convert the effluent into a fine mist. Pneumatic nebulizers have been designed using varied geometries, such as concentric and fixed cross-flow. Other nebulization processes include thermospray, electrospray, and ultrasonic nebulization. Electrospray nebulization has occasionally been paired with CE since it generates a small, highly monodisperse aerosol distribution.

The homogeneity, particle size, and stability of the initial aerosol distribution are important for determining detector sensitivity and reproducibility [13,16,22]. Particularly, nozzle diameter, nebulizer gas pressure, and effluent flow rate have been shown to greatly impact both response and reproducibility [13,23].

In many modern aerosol-based detectors, the largest droplets are removed before the evaporation step to maintain a uniform distribution and reduce evaporation temperatures [4,24]. Several different removal methods are employed, such as a diffuser for Agilent's ELSDs or an impactor in the case of the Corona CAD™, Corona Ultra™, and Alltech ELSDs. SEDERE's design involves condensing the largest droplets on the walls of a glass nebulization chamber shown in Figure 1.2 [24]. The drawback to all these approaches is a large proportion of the effluent, as well as analyte, is diverted to waste (in some cases, > 90%). Nevertheless, such a process is necessary to make these detectors compatible with conventional flow rates and reversed-phase chromatography [5].

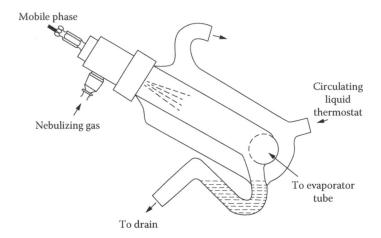

Mobile phase

Nebulizing gas

Circulating
liquid
thermostat

To evaporator
tube

To drain

FIGURE 1.2 Nebulization chamber for SEDEX ELSD designed to condense large aerosol droplets and divert to waste before the evaporator drift tube. (Reprinted with permission from M. Dreux, M. Lafosse, and L. Morin-Allory, *LC GC Int.*, 9: 148, 1996.)

After nebulization, the ELSD, CAD, and CNLSD all employ an evaporation step to remove eluent, thereby generating a dried, tertiary aerosol of solute particles. The size of these particles D is given by the equation

$$D = D_0 \left(\frac{c}{\rho} \right)^{1/3}$$
(1.1)

where c is concentration, D_0 is droplet diameter, and ρ is droplet density.

In the CL aerosol detector, the solvent removal process is unnecessary since the primary wet aerosol deposits solute directly by impacting with a heated, porous catalyst where CL occurs [20]. Following nebulization and evaporation, the mechanics of the different aerosol-based detectors diverge and are discussed in detail in subsequent sections.

1.3.1 EVAPORATIVE LIGHT-SCATTERING DETECTOR

After evaporation, the dried tertiary aerosol particles containing solute pass through a beam of light in the visible range (390 to 700 nm) produced by either a polychromatic source (e.g., halogen lamp) or monochromatically by a laser. Scattered light intensity is then measured with a photodetector, such as a photomultiplier tube (PMT) or photodiode. It should be mentioned that solute molecules with a visible chromophore may absorb the incident light and exhibit greatly diminished ELSD sensitivity when monochromatic light sources are used [25,26].

Three scattering regimes relevant to ELS detection are Rayleigh, Mie, and geometric scattering (sometimes referred to as reflection–refraction) [12]. Rayleigh scattering occurs at small diameters, $D/\lambda < 0.1$, where D is the dried aerosol particle diameter and λ is the wavelength of incident light. This translates to particles less than approximately 100 nm. Mie scattering occurs for intermediate-diameter particles, and geometric scattering occurs for the largest particles, $D/\lambda > 10$.

Scattering intensity I can be modeled by the following equation from La Mer [27–29]:

$$I = kND^2 \left(\frac{D}{\lambda} \right)^y \qquad (1.2)$$

where k is a constant, N is the number of particles, and y depends on the scattering regime (y can be calculated by Mie theory and decreases with D from 4.0 for Rayleigh scattering to –2.2) [30,31]. The intensity of Mie scattering is much greater than Rayleigh, as shown in Figure 1.3, with the optimum particle size for maximum sensitivity occurring at about 380 nm. This imposes a severe constraint on minimizing detection limits since particle size decreases as the cube root of concentration per Equation (1.1). Practically, this corresponds to minimum detection limits of 0.1 to 1 µg/mL (or 1–10 ng on column) for HPLC with conventional bore columns. Microcolumn (0.3- to 1.0-mm inside diameter [I.D.] columns) chromatography has demonstrated lower mass detection limits of about 0.5 ng on column [32].

The intensity of different types of scattered light exhibits an angular dependence [4,33]. While forward scattering is most intense, limiting optical background interference requires that the photodetector be positioned at an angle between 90° and 140° relative to the incident light.

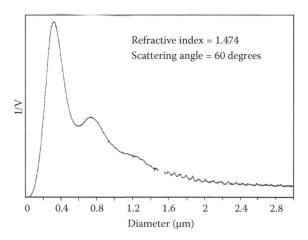

Refractive index = 1.474
Scattering angle = 60 degrees

I/V

0   0.4   0.8   1.2   1.6   2   2.4   2.8

Diameter (µm)

FIGURE 1.3   Scattered intensity per volume of a soybean oil aerosol. (Reprinted [adapted] with permission from P. Vandermeeren, J. Vanderdeelen, and L. Baert, *Anal. Chem.*, 64: 1056, 1992. Copyright 1992 American Chemical Society.)

1.3.2 Condensation Nucleation Light-Scattering Detector

The CNLSD arose from the need to overcome the physical limitation imposed by inefficient Rayleigh scattering for particles smaller than 100 nm [14]. Koropchak and coworkers modified a technology that had been established in atmospheric chemistry, the condensation particle counter (CPC), and applied it as a detector for HPLC, size exclusion chromatography (SEC), CE, capillary electrochromatography (CEC), and ion chromatography (IC) separations [15,35–40]. Jorgenson et al. also found success applying the CPC to CE using an electrospray interface where they were able to count individual macromolecules [41]. By using the dried aerosol particles as nucleation sites for liquid condensation, particles from 2 to 3 nm could be grown up to 10 μm [42,43].

As previously mentioned, nebulization and eluent evaporation are also required steps for CNLSD. After these, a condenser is also needed to remove solvent vapor. Dried aerosol then enters a growth chamber, where the condensing solvent vapor pressure is carefully controlled so only heterogeneous nucleation is initiated [14]. Homogeneous nucleation, which results in increased background noise, occurs when the saturation solvent spontaneously forms droplets. Butanol was initially used as a condensing fluid, while the commercially available CNLSD from Quant Technologies uses water.

A difficulty encountered when a CPC was first paired with chromatographic instruments was moderately high background noise due to residual nonvolatile contaminants from organic solvents or column bleed [15]. To overcome this, diffusion screens were added before the saturator to remove very small particles [15]. Improvements in HPLC detection limits versus ELSD were found to be several orders of magnitude, and the detector response versus concentration was found to be more linear for the CNLSD [44].

1.3.3 Charged Aerosol Detector

Similar to CNLSD, the CAD had its origins in detection methods previously applied to the analysis of aerosols [45]. Unlike light scattering, aerosol charging can be applied efficiently to particles smaller than 100 nm, resulting in improved detection limits compared to ELSD [16]. In fact, maximum sensitivity per unit particle mass S_m occurs for particles with about a 10-nm diameter according to the following set of equations:

$$S_m = \frac{4.4 \times 10^5}{\rho} D^{3.6} \qquad \text{for } D < 10 \text{ nm}$$

$$S_m = \frac{3.01 \times 10^{11}}{\rho} D^{-1.89} \qquad \text{for } D > 10 \text{ nm} \qquad (1.3)$$

Nebulization and evaporation steps are also required for the CAD. A stream of dried, purified N_2 gas becomes positively charged after being ionized by a high-voltage corona discharge needle. The positively charged N_2 gas then transfers its charge after

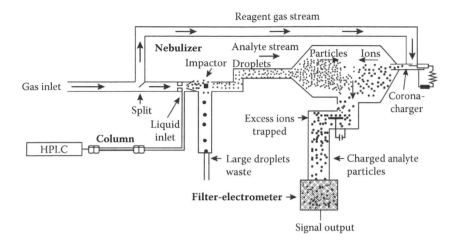

FIGURE 1.4 Corona charged aerosol detector (CAD) schematic. (Reprinted with permission from P. H. Gamache, R. S. McCarthy, S. M. Freeto, D. J. Asa, M. J. Woodcock, K. Laws, and R. O. Cole, *LC GC Eur.*, 18: 345, 2005.)

mixing with the dried aerosol particles. The charge on each aerosol particle is related to its size, which is proportional to concentration. An electrometer then measures the flux from these particles as they collect on a filter. A detector schematic is presented as Figure 1.4.

It is important to recognize the difference in the CAD's operating principles versus MS ion sources, such as atmospheric pressure chemical ionization (APCI). In APCI, a heater (often up to 500°C) is used to rapidly vaporize analyte molecules. The gas phase analyte *molecules* then become ionized after reaction with positively charged reagent gas. For the CAD, the analyte is an aerosol *particle* in the solid phase, and charge is transferred to this solid particle after mixing with reagent gas (i.e., N_2^{+*}). High-mobility ions, such as those that would be in the gas phase from an APCI-type process or excess N_2^{+*}, are removed via a negative ion trap. Charged aerosol particles are low mobility. Thus, they are unaffected by the ion trap and can be accurately measured by a particle filter connected to an electrometer.

1.3.4 AEROSOL CHEMILUMINESCENT DETECTOR

Chemiluminescence on the surface of a solid is an interesting phenomenon. Breysse et al. reported that carbon monoxide can undergo catalytic oxidation on the nonporous surface of thoria, which creates a weak CL emission [46]. This phenomenon of catalytic oxidation of organic vapors was used to design gas sensors [47,48].

Zhang et al. developed a prototype CLAD for HPLC analysis. Their design employed porous alumina, which is a material having a large surface area, stable structure, good adsorption, and high activity [20]. The detector schematic is provided as Figure 1.5. CLAD functions by converting the HPLC effluent into an aerosol via a nebulizer.

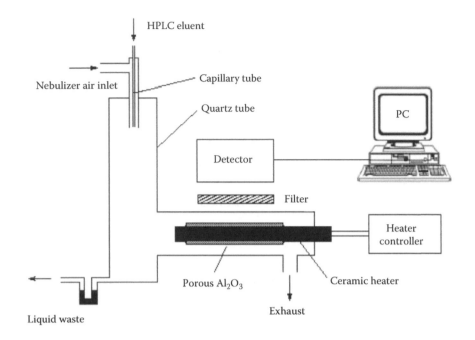

HPLC eluent

Nebulizer air inlet

Capillary tube

Quartz tube

Detector

PC

Filter

Heater controller

Porous Al$_2$O$_3$

Ceramic heater

Liquid waste

Exhaust

FIGURE 1.5 Schematic diagram of the chemiluminescence aerosol detector (CLAD). (Reprinted [adapted] with permission from Y. Lv, S. C. Zhang, G. H. Liu, M. W. Huang, and X. R. Zhang, *Anal. Chem.*, 77: 1518, 2005. Copyright 2005 American Chemical Society.)

The aerosol is then deposited on the porous alumina, which has been immobilized on a ceramic tube. This tube can be heated to provide an optimized temperature for CL to occur. Finally, the light emitted by CL is detected by a computerized ultraweak CL analyzer equipped with a PMT.

Application of the CLAD for saccharide analysis was studied. The CL spectra of sucrose, glucose, maltose, R-lactose, and raffinose from reaction on the surface of porous alumina at different temperatures were collected. Since similar spectra with peaks at 425 and 460 nm were observed, it was proposed that CL detection may be due to C_2H_2 catalytic oxidation induced by alumina. Accordingly, a CL wavelength of 460 nm was chosen to study detection limits and linearity. It was shown that the response for each saccharide was linear in the concentration range from 10 to 100 μg/mL, and the limit of detection was as low as 3.1 μg/mL for glucose. The effect of nebulizer gas flow rate and temperature on CL response was investigated, and it was found that optimum conditions were an airflow rate of 14 L/min and temperature of 400°C. Finally, the authors analyzed glucose and sucrose by coupling HPLC with the CLAD. Different acetonitrile-to-water ratios were used, and at 75:25 an optimum separation was achieved with baseline noise relatively constant for different acetonitrile amounts.

Unlike other aerosol-based detectors, for which only nonvolatile or semivolatile compounds can be detected, CLAD can be used for volatile analytes. As shown in

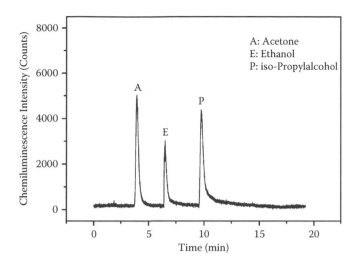

FIGURE 1.6 CLAD chromatogram for volatile compounds. A, acetone (100 µg/mL); E, ethanol (500 µg/mL); P, isopropyl alcohol (200 µg/mL); 1.0 mL/min carrier gas flow rate; 12 L/min nebulizer airflow rate; 400°C catalyst temperature; 460-nm wavelength. (Reprinted [adapted] with permission from Y. Lv, S. C. Zhang, G. H. Liu, M. W. Huang, and X. R. Zhang, *Anal. Chem.*, 77: 1518, 2005. Copyright 2005 American Chemical Society.)

Figure 1.6, ethanol, acetone, and propanol were directly observed with water as a mobile phase.

The authors also examined the compatibility of the CLAD with common reversed-phase solvents, such as acetonitrile and methanol. It was shown that greater than 10% by volume methanol in water could induce CL, making this solvent incompatible with the technique. Fortunately, CL was not observed for acetonitrile at any percentage in water.

The other major advantage of the CLAD was insensitivity to inorganic salts since only volatile buffers are compatible with the other aerosol-based detectors. Zhang et al. separated amino acids L-histidine and L-phenylalanine using sodium phosphate buffers (NaH_2PO_4/Na_2HPO_4, pH 7.4) in concentrations ranging from 0 to 0.1 M, and they observed no significant change in response.

Zhang and coworkers also demonstrated the compatibility of the CLAD when paired with CE [21]. Saccharides were chosen to study the effect of buffer pH and concentration on response. In contrast to electrochemical detection, the CL response was insensitive throughout the pH and buffer range from 8 to 12 and 25 to 75 mM, respectively. These results demonstrated an advantage for the CLAD over the electrochemical detector since pH and buffer concentration can be fully used to optimize separations. Due to the low flow rates typical of CE, a sheath liquid was necessary to provide stable aerosol formation. Figures of merit for analysis of seven saccharides were linear response from 30 to 2000 µg/mL, percent relative standard deviation (%RSD) from 2.1% to 4.1% for 11 replicates, and detection limits from 10 to 20 µg/mL, which is slightly better than indirect UV detection.

1.4   COMPARISON OF PERFORMANCE

Several companies manufacture and sell ELSDs: SEDERE, Agilent Technologies, Waters Corporation, Grace Davison Discovery Sciences, Teledyne ISCO, and Yamazen Science. Shimadzu sells a branded version (ELSD-LTII) of SEDERE's 85LT for use with its instruments. SEDERE's LT90 and Agilent's 1290 Infinity ELSD both utilize short-wavelength 405-nm "blue" laser light sources, which provide higher scattering intensity for small particles [49], resulting in lower detection limits versus older model ELSDs. SEDERE's LT90 ELSD allows for detection of semivolatiles and can handle the widest range of flow rates, from those typical of CE (5 µL/min) up to those of preparative chromatography (5 mL/min after splitter). Agilent's ELSD has an interesting "real-time gas programming" feature that Agilent reports can minimize gradient solvent effects. Waters's ELSD features a small footprint, fits conveniently with other ultra performance liquid chromatography (UPLC®) components, and is directly controlled by Waters's Empower™ software. Grace claims low-nanogram detection limits and good performance for semivolatiles from their Alltech® 3300 ELSD. Teledyne ISCO markets a system, CombiFlash Rf 200i, that integrates an ELSD and UV-visible detector with flash chromatography; in addition, it sells stand-alone ELSDs for both analytical and preparative chromatography. Yamazen also sells a system, ELSD-100X, designed for flash chromatography.

Dionex (now a part of Thermo Scientific) manufactures and sells the Corona CAD™ and Corona Ultra RS™ detectors. The Corona CAD was the first commercially available CAD for HPLC applications, while the Corona Ultra was designed specifically to handle fast acquisition rates and minimal band broadening required for ultra high-performance liquid chromatography (UHPLC).

Quant Technologies currently manufactures and markets the only available CNLSD, referred to as the NQAD. This comes in three models: QT-500, QT-600, and QT-700. The QT-600 was specifically designed for analysis of semivolatiles as it allows for subambient mobile phase evaporation, while the QT-700 adds the features of analytical-scale fraction collection and solvent recycling.

A comparison between the specifications for several state-of-the-art aerosol-based detectors in each class is provided in Table 1.1. Most detectors now feature dynamic ranges of four orders of magnitude, are UHPLC compatible, and have special controls to quantitate semivolatiles, either by improved nebulizer design or subambient solvent evaporation.

Table 1.2 summarizes the findings from several studies in recent years on the various aerosol-based detectors. Generally, detection limits are about one or two orders of magnitude higher for the ELSD versus the CNLSD and CAD. The lowest detection limits (subnanogram) on all three detector types were found by Hutchinson and coworkers when using UHPLC with a narrow-bore column [50]. They compared four detectors for use with UHPLC: NQAD QT-500, Corona CAD, Corona Ultra, and Varian's 385-LC ELSD. The CNLSD provided the lowest detection limits, while the Corona CAD exhibited the best precision. Other recent studies that have attempted direct comparison between different types of aerosol-based detectors have found lower detection limits and better precision for the CAD versus ELSDs [51–55]. In addition, the response curve for the CAD has generally been found to be more

TABLE 1.1

Comparison of Aerosol-Based Detector Specifications

Detector	ELSD	ELSD	CAD	CNLSD
Manufacturer	SEDERE	Agilent	Thermo Scientific	Quant Tech.
Model	LT90	1290 Infinity	Corona™ ultra RS™	NQAD QT600
Flow rate (mL/min)	0.005–5[a]	0.2–5	0.2–2.0	0.1–2.2
Mobile phase compatibility	Volatile buffers	Volatile buffers	Volatile buffers (pH ≤ 7.5)	Volatile buffers
Dimensions (w × d × h)	10 × 22 × 18 in.	8 × 18 × 16 in.	17.5 × 22 × 9 in.	7 × 17 × 12.5 in.
Weight	40 lb	29 lb	27 lb	25 lb
Evaporation temperature range	Ambient: 100°C	10–80°C	Ambient	10–60°C
Nebulizer temperature control	Not controlled	25–90°C	5–35°C	15–40°C
Maximum acquisition rate	100 Hz	80 Hz	200 Hz	100 Hz
UHPLC compatible?	Yes	Yes	Yes	Yes
Gradient compatible?	Yes	Yes	Yes	Yes
Linear response?	No[b]	No[b]	No[b]	No[b]
Dynamic range	10^4	Not available	10^4	10^4

[a] Different nebulizers required depending on flow rate and application (e.g., HPLC, UHPLC, or SFC).

[b] Linear fits are reasonable for short ranges of typically one to two orders of magnitude.

linear over short ranges of one or two orders of magnitude than both the ELSD and CNLSD [53,55–57]. However, Shaodong et al. actually found the ELSD provided a more linear response than the CAD [54]. A word of caution regarding direct detector comparisons: Performance characteristics are highly dependent on the particular type of analyte, LC conditions, as well as differences in detector design.

1.5 LIMITATIONS

Megoulas and Koupparis, in their review of the ELSD, cited many limitations of this technology that are also applicable to other aerosol-based detectors [5]. These limitations include a requirement for volatile mobile phases, sensitivity to column bleed (and nonvolatile mobile phase contaminants), nonlinear response, moderate detection limits, low sensitivity for volatile and thermally labile compounds, sample destruction, nonselectivity, and little ability to provide information regarding compound identification (as opposed to MS and spectroscopic techniques). Several of these limitations are quite serious and are given more attention in the next few sections. The moderate detection limits exhibited by ELSD (especially older-model ELSDs) have been improved on by newer aerosol-based detectors, such as the CAD and CNLSD. The issue of nonselectivity is often viewed as a significant advantage for this class

TABLE 1.2
Figures of Merit from Recent LC Studies Using Aerosol-Based Detectors

Detector	Model	Separation Mode[d]	Analytes	Precision (%RSD)		LODs[d] (ng)	Reference
				Major Component[a]	Minor Component[a]		
ELSD	Alltech 2000ES	RP	Fatty acids[b]	0.7 to 4.7	1.1 to 11.2	14 to 51	51
ELSD	ESA ELSD	NP	Phospholipids	1.9 to 10.7	3.4 to 17.5	71 to 1195	52
ELSD	SEDERE Sedex 75	HILIC	Carbohydrates	N/A	2.4 to 7.6	N/A	58
ELSD	Varian 385-LC ELSD	RP	Pharmaceuticals[b]	4.5 to 21.5	N/A	0.5 to 24[c]	50
ELSD	SEDERE Sedex 75	RP	Saikosaponin	N/A	1.4 to 19.2	30 to 90	53
ELSD	Alltech 2000	RP	Pharmaceuticals	2.6 to 8.3	4.3 to 26.6	73 to 77	54
ELSD	SEDERE Sedex 75	RP	Ginsenosides	≤4.0	N/A	24 to 60	55
ELSD	Alltech 800	HILIC	Pharmaceutical	0.4	N/A	N/A	59
ELSD	Waters 2424	NP	Oxidized FAMEs[d]	2.5 to 4.5	N/A	50 to 232	60
ELSD	SEDERE Sedex 85	RP	Kryptofix 2.2.2	1.4 to 2.4	0.8 to 2.3	20	61
ELSD	Alltech ELSD	RP	Saponins	1.6 to 4.8	N/A	8 to 13	62
ELSD	Alltech 2000	NP	Lipids	0.3 to 14.7	N/A	N/A	63
CNLSD	NQAD QT-500	RP	Polymers	4.1 to 5.5	N/A	N/A	64
CNLSD	NQAD QT-500	HILIC	Amines[b]	0.8 to 1.9	N/A	4 to 27	56
CNLSD	NQAD QT-500	RP	Pharmaceuticals[b]	1.3 to 30.0	N/A	0.01 to 0.4[c]	50
CNLSD	NQAD QT-500	RP	Antibiotics	N/A	N/A	15 to 27[c]	65
CAD	Corona CAD	RP & HILIC	Glycopeptides	1.8	N/A	0.9 to 2	66
CAD	Corona Plus	RP	Fatty acids[b]	0.4 to 2.2	0.6 to 1.7	4 to 11	51
CAD	Corona CAD	NP	Phospholipids	1.9 to 9.4	2.9 to 11.9	15 to 249	52
CAD	Corona Plus	RP	Lipids	N/A	0.5 to 3.7	2 to 4	67

CAD	Corona CAD	NARP	Triacylglycerols	0.4	1.3 to 5.8	0.4 to 2	68
CAD	Corona CAD	HILIC	Amines[b]	0.4 to 2.1	N/A	1 to 10	56
CAD	Corona CAD	RP	Pharmaceuticals[b]	0.9 to 9.7	N/A	0.08 to 0.4[c]	50
CAD	Corona Ultra	RP	Pharmaceuticals[b]	1.2 to 18.6	N/A	0.3 to 1[c]	50
CAD	Corona CAD	HILIC	Carbohydrates	N/A	N/A	3 to 18	69
CAD	Corona Plus	Mixed mode	Counterions[b]	0.7 to 1.1	N/A	0.5 to 400	70
CAD	Corona CAD	RP	Saikosaponin	N/A	1.7 to 6.0	5 to 15	53
CAD	Corona Plus	RP	Pharmaceuticals	1.9 to 3.9	6.2 to 9.5	16 to 48	54
CAD	Corona CAD	RP	Ginsenosides	≤2.9	N/A	11 to 40	55

[a] Minor component precision for concentrations < about 20 μg/mL; major component precision for concentrations > about 20 μg/mL.

[b] Known volatiles and semivolatiles included in test set, which may have higher detection limits and RSDs.

[c] UHPLC analysis using 2.1-mm I.D. column.

[d] RP = reversed phase; NP = normal phase; HILIC = hydrophilic liquid interaction chromatography; NARP = nonaqueous reversed phase; LOD = limit of detection; FAMEs = fatty acid methyl esters.

TABLE 1.3
pH Range of Volatile Buffers

Buffer	Recommended pH Range
Ammonium-formate	2.8–4.8 and 8.2–10.2 [b]
Ammonium-acetate	3.8–5.8 and 8.2–10.2 [b]
Ammonium-bicarbonate[a]	6.8–11.3
Triethylammonium-acetate	10.0–12.0

[a] Lower pH using acetic acid.

[b] Acetate or formate is a recommended counterion for buffering at ammonium's pK_a.

of detectors since they respond to most compounds. Last, sample destruction is not a major detriment due to the small injection volumes characteristic of LC and CE.

1.5.1 MOBILE PHASE COMPATIBILITY

Volatile, high-purity mobile phases are necessary for compatibility. The CNLSD is particularly sensitive to nonvolatile mobile phase impurities for reasons previously discussed [15]. For low-level quantitation, it is recommended that LC-MS-grade solvents be used. Mobile phase additives must also be volatile. Examples of commonly used volatile additives are formic acid, acetic acid, trifluoroacetic acid (TFA), ammonium bicarbonate [71], ammonium hydroxide, and triethylamine (TEA). Only volatile buffers can be used. Table 1.3 provides the pH range of recommended volatile buffers. It is generally advantageous to use the lowest buffer concentrations that provide adequate peak shapes, especially for reversed-phase chromatography, to further reduce noise. Another best practice is to thoroughly rinse mobile phase bottles with both water and organic solvents before use to remove any residual cleaning agents.

Some combinations of volatile acids and bases can form nonvolatile salts, such as ammonium-trifluoroacetate, and consequently should be avoided [72]. When switching between mobile phases containing fluorinated acids and those containing volatile amines, it is recommended to rinse the lines and detector with pure water to remove residual salts. This helps ensure low background noise.

At the time of this writing, one disadvantage of the CAD is higher background noise at mobile phase pH greater than 7 compared to the ELSD and CNLSD. In particular, ammonium-bicarbonate buffers are not recommended for use with this detector.

1.5.2 COLUMN COMPATIBILITY

Similar to MS, it is important for compatibility to use columns with low bleed to minimize background noise. For reversed-phase and hydrophilic interaction liquid chromatography (HILIC) separations at neutral to high pH using silica-based stationary phases, this is especially important. Even though most silica columns

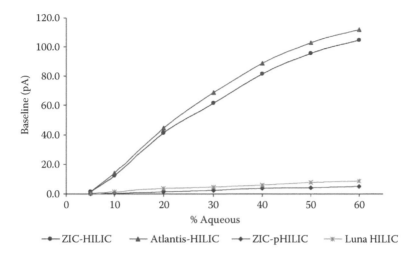

FIGURE 1.7 (See Color Insert.) Column bleed from four different HILIC columns as a function of aqueous component of mobile phase. Detector = Corona® CAD; column temperature = ambient; mobile phase buffer = 0.1 M ammonium acetate, pH 7.0; organic solvent = acetonitrile; flow rate = 1.0 mL/min. (Reprinted from Z. Huang, M. A. Richards, Y. Zha, R. Francis, R. Lozano, and J. Ruan, *J. Pharm. Biomed. Anal.*, 50: 809, 2009, with permission from Elsevier. Copyright 2009.)

list pH compatibility from 2 to 8, they may exhibit significantly higher noise near the limits of their range, as shown in Figure 1.7. At pH 7.0, the polymeric zic-*p*HILIC column displayed minimal bleed, while the same bonded chemistry with a silica-based support, zic-HILIC, had baseline noise greater than 10 times as high.

Several bonded phases, such as amino and cyano, that are frequently used for separations with UV detection may exhibit surprisingly high background noise throughout their recommended pH ranges. In these instances, columns with polymeric supports can be useful alternatives. For instance, Asahipak NH2P-50 4E is a polymeric amino column known to exhibit low bleed and good compatibility with aerosol-based detectors [58].

Teutenberg et al. studied the effect of high-temperature LC (up to 200°C) on column degradation using both a UV detector and a CAD [74]. The CAD, being much more sensitive to bleed, was a better indicator of degradation. From Figure 1.8, the noise was significantly higher at elevated temperature for a silica-based column, Luna C18, than columns specially designed to withstand extreme conditions. Of all the columns tested, a novel column from ZirChrom using graphite on a TiO$_2$ support exhibited the lowest bleed.

1.5.3 LINEARITY

All aerosol-based detectors exhibit a nonlinear response over large concentration ranges, and this is a major limitation for these detectors seeing greater use. Deviations from

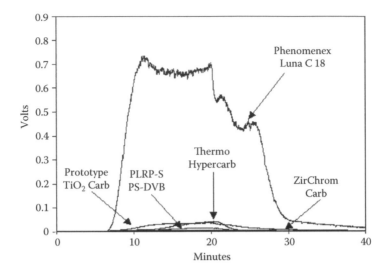

FIGURE 1.8 Baseline noise for five HPLC columns. Detector = Corona CAD; mobile phase = water; flow rate = 0.5 mL/min; temperature gradient = hold at 30°C for 5 min, then 30 to 200°C in 5 min, followed by holding at 200°C for 10 min. PS-DVB = polysulfonated divinylbenzene; PLRP-S = polymeric reversed phase. (Reprinted from T. Teutenberg, J. Tuerk, M. Holzhauser, and T. K. Kiffmeyer, *J. Chromatogr. A*, 1119: 197, 2006, with permission from Elsevier. Copyright 2006.)

linearity are less dramatic for the CAD and the CNLSD, but these become more apparent over broad ranges and at different eluent compositions.

By using Mie scattering theory, Mourey and Oppenheimer accurately modeled the ELSD response curve over two orders of magnitude as shown in Figure 1.9 [13]. At the high end of the range, the PMT became saturated, causing a deviation from theory. Vandermeeren et al. later were able to predict the efficiency loss due to PMT saturation [34]. From these curves, it is obvious that ELSD response is nonlinear and sigmoidal over large ranges. This behavior is due primarily to changes in light-scattering efficiency as a function of aerosol particle size, which was shown in Figure 1.3. At low concentrations where Rayleigh light scattering occurs, sensitivity drops dramatically due to much lower scattering efficiency, resulting in a characteristic "boot" shape to ELSD response curves.

Guiochon et al. applied the following function to describe an ELSD's calibration curve [22]:

$$A = a(M^b) \tag{1.4}$$

where A is detector response (in this context, peak area), M is mass, and a and b are constants. By taking the base 10 logarithm of both sides of Equation (1.4), the function can be linearized, with a slope of b and y-intercept of $\log a$.

$$\log A = \log a + b \times \log M \tag{1.5}$$

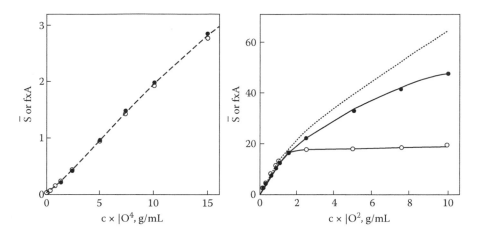

FIGURE 1.9 ELSD calibration curve for polystyrene (concentration, c = 10⁻⁸); f, scaling factor; A, peak area; 20-psi nebulizer gas pressure; 60°C evaporator temperature; (o) normal photomultiplier voltage; (•) reduced photomultiplier voltage; (...) Mie scattering model prediction. (Reprinted [adapted] with permission from T. H. Mourey and L. E. Oppenheimer, *Anal. Chem.*, 56: 2427, 1984. Copyright 1984 American Chemical Society.)

This treatment is valid over several orders of magnitude for the ELSD. It cannot be used to model the entire curve due to the sigmoidal behavior from multiple scattering regimes evident from Figure 1.9. The value of *b* depends primarily on the range of the calibration curve, the RI of the dried aerosol particles, and characteristics of the particular ELSD design [33]. This coefficient is often measured between 0.9 and 1.8 [24], and it has been observed that structurally similar compounds exhibit nearly the same values due to similar RIs [75].

Using chemometrics, Krull et al. found that minimally six standard levels equally spaced across a double-log calibration curve (range about two orders of magnitude) provided accurate results for ELSD data [10].

Koropchak et al. noted that the CNLSD provided a more linear response than the ELSD since in theory all particles are grown to approximately the same size [14,44]. However, Figure 1.10 shows a calibration curve for the CNLSD, displaying sigmoidal shape for a range of 20 to 180 µg/mL for polysorbate 20 [64]. The reason for the decreased sensitivity at high concentration may be due to PMT saturation or to another scattering regime (such as refraction-reflection) dominating. Nevertheless, the authors applied a linear fit over the concentration ranges 10–60 µg/mL for polysorbate 20 and 2–40 µg/mL for polyethylene glycol (PEG), and they obtained reasonable correlation coefficients, $r^2 > 0.99$.

For the CAD, Equation (1.4) is applicable over a much larger range of roughly four orders of magnitude. Moreover, a linear fit is reasonable when applied to this detector for ranges of up to two orders of magnitude. Figure 1.11 shows typical response curves for the CAD for a range from 10 to 1000 µg/mL.

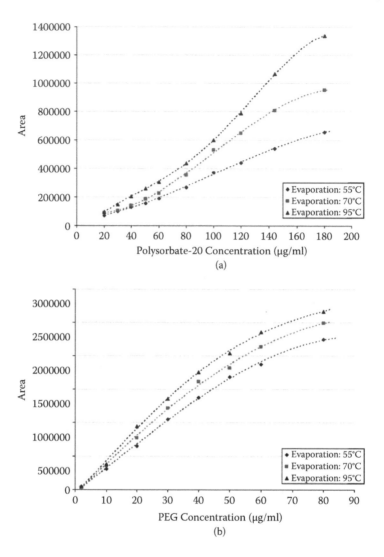

FIGURE 1.10 CNLSD response versus concentration curves at three evaporation temperatures (55°C, 70°C, and 95°C) for (A) polysorbate 20 and (B) PEG. Column = kinetex C18 (2 μm, 100 × 3 mm); mobile phase A = methanol/water/trifluoroacetic acid (10:90:0.1, v/v/v); mobile phase B = methanol/water/trifluoroacetic acid (90:10:0.1, v/v/v); flow rate = 0.6 mL/min; gradient = 83% to 100% B in 4.5 min, then 2 min hold; injection volume = 5 μL. (Reprinted from S. Fekete, K. Ganzler, and J. Fekete, *J. Chromatogr. A*, 1217: 6258, 2010, with permission from Elsevier. Copyright 2010.)

The nonlinear response of aerosol-based detectors has an important implication for purity determinations when area percentage (also known as area normalization) quantitation is used [77]. In most cases, an ELSD and a CNLSD will both underestimate impurity levels, whereas the CAD will overestimate them due to the shape of their respective response curves (Figure 1.9, Figure 1.10, and Figure 1.11). The level

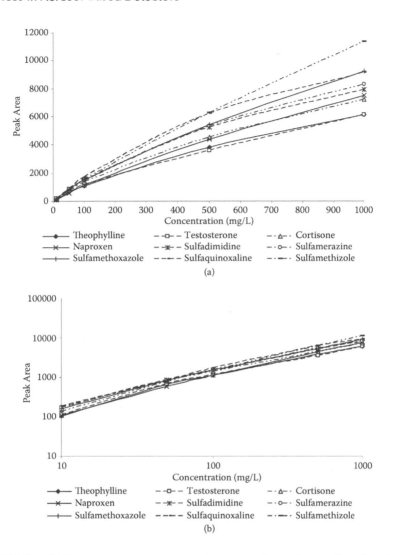

FIGURE 1.11 CAD response versus concentration curves for a mixture of small-molecule pharmaceuticals from an isocratic SFC separation: (a) linear scale; (b) logarithmic scale; mobile phase = methanol/CO_2 (8:2, v/v); column = ethylpyridine; column temperature = 40°C; flow rate = 2.0 mL/min; outlet pressure = 100 bar. (Reprinted [adapted] with permission from C. Brunelli, T. Gorecki, Y. Zhao, and P. Sandra, *Anal. Chem.*, 79: 2472, 2007. Copyright 2007 American Chemical Society.)

of the deviation can be quite significant. For the ELSD with a typical organic compound's RI and pure Mie scattering, the *b* term in Equation (1.4) is 1.33 [78]. In this case, a 10% w/w impurity would be measured as 5% w/w. The reason for this error can be understood by the following example: Consider a sample with the major component designated as compound 1 and impurity designated as compound 2, having masses M_1 and M_2 and peak areas A_1 and A_2, respectively. To simplify calculations, these compounds have identical a and b terms per Equation (1.5). The calibration curves are then

$$\log A_1 = \log a + b \times \log M_1$$

$$\log A_2 = \log a + b \times \log M_2$$

Subtracting the first equation from the second yields

$$\log(A_2/A_1) = b \times \log(M_2/M_1)$$

Setting $M_1 + M_2 = 100$ and $A_1 + A_2 = 100$, which is the case for area normalization, results in the following equation after algebraic rearrangement [77]:

$$A_2 = \frac{100 \left[\dfrac{M_2}{100 - M_2} \right]^b}{\left\{ 1 + \left[\dfrac{M_2}{100 - M_2} \right]^b \right\}} \tag{1.6}$$

Figure 1.12 is a plot of Equation (1.6) for different values of b showing the error relative to the true weight/weight percentage of the impurity. Note that this error is amplified as the impurity level decreases. The CAD will exhibit b values less than 1

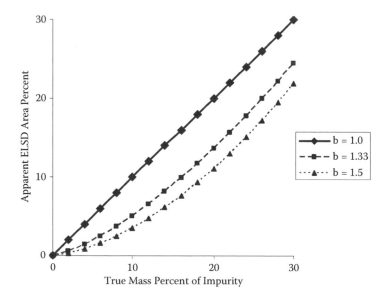

FIGURE 1.12 Deviation of area percentage for impurity determination by ELSD from actual weight/weight percentage. (Reprinted [adapted] with permission from S. Lane, B. Boughtflower, I. Mutton, C. Paterson, D. Farrant, N. Taylor, Z. Blaxill, C. Carmody, and P. Borman, *Anal. Chem.*, 77: 4354, 2005. Copyright 2005 American Chemical Society.)

and will therefore overestimate impurity levels when area percentage quantitation is used. Therefore, to obtain accurate purity determinations with an aerosol-based detector, individual calibration needs to be done for each peak. Note that this can still be performed via a single calibration curve (i.e., universal calibration), but it rules out the possibility of using a simple area percentage or area ratio calculation as is often done with UV detection.

1.5.4 VOLATILITY LIMIT

In principle, only analytes with a lower volatility than the eluent can be quantified with aerosol-based detection. In practice, we have found that an analyte's volatility must be sufficiently lower than this. The purposes of this section are to examine the practical limit to quantitation of semivolatiles and volatiles and to discuss methods to improve it.

Consider Figure 1.13 for the separation of aspirin, acetylsalicylic acid, and its potential impurities under isocratic reversed-phase elution. The aspirin peak and major impurity/degradant peak, salicylic acid, were not detected by CAD, whereas the larger molecular weight impurities were detected. All these compounds have much lower vapor pressures than any of the mobile phase constituents. Note that all these compounds were injected at the relatively low concentration of 4 μg/mL; however, this concentration is typical of impurity levels. At higher concentrations, the aspirin and salicylic acid peaks would be detected, albeit with greatly diminished responses relative to "nonvolatiles."

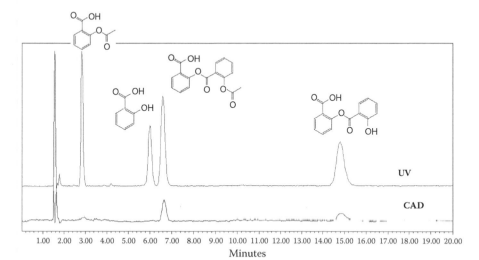

FIGURE 1.13 Isocratic elution of aspirin and its impurities at approximately equal concentration, 4 μg/mL. Mobile phase = methanol/tetrahydrofuran/water/trifluoroacetic acid (44:5:51:0.1, v/v/v/v); column = xBridge C18; column temperature = 30°C; flow rate = 1.0 mL/min; injection volume = 10 μL.

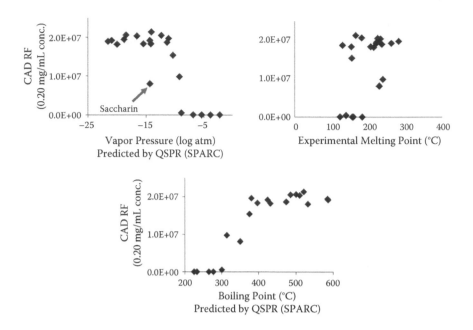

FIGURE 1.14 Plots between CAD response factors and physical properties for a diverse set of small molecule drug substances. Vapor pressure and boiling point predicted using QSPR package from University of Georgia known as SPARC (http://ibmlc2.chem.uga.edu/sparc/) [80].

This example highlights a significant problem for all aerosol-based detectors. It would be extremely beneficial if there were a way to predict a compound's detectability based solely on its molecular structure. Several researchers have found success correlating a compound's response with molecular properties obtained via quantitative structure property relationships (QSPRs) [77,79]. Aerosol-based detectors are known to provide a uniform response under isocratic elution (discussed in more detail in subsequent sections). Figure 1.14 shows the plot of CAD response factors versus molecular properties for 20 small-molecule compounds (molecular weights less than 1000) with diverse structures. From these plots, predicted boiling point exhibited the strongest correlation with response.

The relationship between predicted boiling point and CAD response is expanded in Figure 1.15 [80], showing distinct regions. Volatile compounds that were not detected even at relatively high concentration had predicted boiling points less than 300°C. Semivolatile compounds, which were detected but had diminished response, had predicted boiling points from 300 to 400°C. Nonvolatiles had predicted points greater than 400°C. Note that boiling point prediction implies that a compound would boil at that temperature and atmospheric pressure in the *absence of degradation*. Most organic molecules evaluated will degrade at temperatures much lower than their predicted boiling point. Thus, this prediction is really only of value to determine if a compound will be detectable by the CAD. Other aerosol-based detectors may

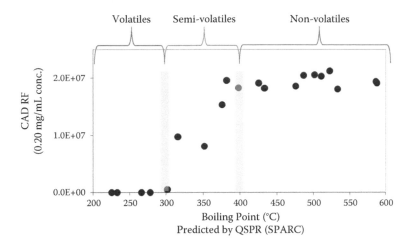

FIGURE 1.15 Predicted boiling point versus CAD response factor showing distinct regions for non-volatiles, semi-volatiles, and volatiles. Boiling points predicted using QSPR package from University of Georgia known as SPARC (http://ibmlc2.chem.uga.edu/sparc/) [80].

exhibit higher or lower limits at which compounds would be considered volatile and semivolatile. Nevertheless, Lane et al. found a similar limit at which compounds are sufficiently nonvolatile when using the Polymer Lab's model 2100 ELSD [77].

One method to improve detectability of semivolatiles involves lowering the temperature at which eluent evaporation occurs. This is shown in Figure 1.16 [80].

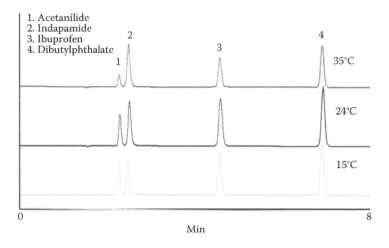

FIGURE 1.16 Subambient evaporation using an ELSD demonstrating improvement in semivolatile (acetanilide and ibuprofen) response. Compounds at approximately equal concentration. (Reprinted from A. W. Squibb, M. R. Taylor, B. L. Parnas, G. Williams, R. Girdler, P. Waghorn, A. G. Wright, and F. S. Pullen, *J. Chromatogr. A*, 1189: 101, 2008, with permission from Elsevier. Copyright 2008.)

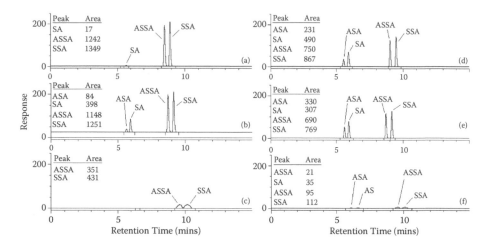

FIGURE 1.17 Effect of buffer and evaporator temperature on NQAD response for aspirin and related compounds. Evaporator temperature of 60°C for chromatograms (a)–(c) and off (~32°C) for chromatograms (d)–(e). Buffers: 1.6 mM NH$_4$-acetate (pH 4.6) for chromatograms (a) and (d), 1.6 mM TMG-acetate (pH 4.6) for chromatograms (b) and (e), and no buffer (100% water) for chromatograms (c) and (f). ASA, acetylsalicylic acid; SA, salicylic acid; ASSA, acetylsalicylsalicylic acid; and SSA, salicylsalicylic acid. All analytes at 0.1 mg/mL in acetonitrile diluent. Conditions: XSELECT CSH C18 column (3.5 μm, 150 mm × 3.0 mm); column temperature 10°C; injection volume 3.0 μL; flow rate 0.5 mL/min. Mobile phase A was buffer, and mobile phase B was acetonitrile; gradient program was 90% to 10% A in 10 min with a 5-min reequilibration time.

There is a trade-off to this approach since nonvolatile compounds generally exhibit greater sensitivity at higher evaporator temperatures, which is why most manufacturers allow the temperature of the drift tube to be thermostatted over a wide range.

Another method to improve detectability involves changing pH of the mobile phase, such that analytes are ionized under the elution conditions. This is demonstrated in Figure 1.17 for the case of aspirin and its impurities. Previously, aspirin and salicylic acid were undetectable at low concentrations when eluted with an acidic mobile phase. By increasing the mobile phase pH to 4.6 using an ammonium acetate buffer, the response of these two semivolatiles was significantly enhanced. Furthermore, by switching to a buffer with a stronger basic component than ammonia, response further improved. This became more apparent as the evaporator temperature was increased from ambient (~32°C) to 60°C. Here, 1,1,3,3-tetramethylguanidine (TMG) was used, which has a pK$_b$ of 0.4 versus 4.8 for ammonia. These results indicate that the composition of the dried aerosol particle is a salt under these conditions. With an ammonium buffer, the salt is an ammonium-acetylsalicylate salt, which would have a lower volatility than an aerosol particle composed of just the free acid of aspirin.

This work was further extended to see if volatile molecules, such as ammonia, methylamine, and hydrazine, could be detected by using acidic mobile phase

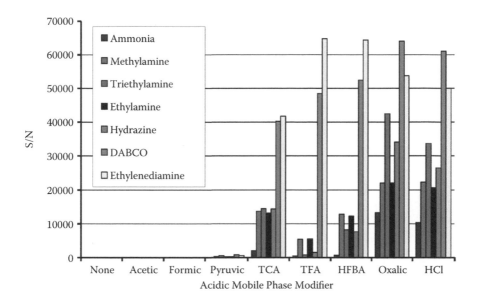

FIGURE 1.18 (See Color Insert.)   Improvement in volatile base response by addition of volatile acidic mobile phase modifier. TCA, trichloroacetic acid; TFA, trifluoroacetic acid; HFBA, heptafluorobutyric acid; DABCO, 1,4-diazabicyclo[2.2.2]octane.

modifiers [56]. Figure 1.18 shows the results from flow injection analysis (FIA) screening of various acidic modifiers, which indicated a correlation between the acid strength of the modifier and response of the basic analyte. When weak acids, such as acetic and formic acid, were used, there was no response for any of the test compounds. When stronger acids, such as hydrochloric and oxalic acid, were used, all analytes had fairly high response. Doubly charged basic analytes ethylenediamine and 1,4-diazabicyclo[2.2.2]octane (DABCO) exhibited the strongest enhancement in sensitivity.

Figure 1.19 shows the separation of basic analytes under isocratic HILIC conditions on a zic-pHILIC column using a mobile phase of TFA/acetonitrile/water (0.04:60:40, v/v/v). The addition of 0.2 mM HCl to this mobile phase enabled the detection of ammonia. Figures of merit for this separation were quite good, with reported detection limits from 1 to 10 ng on column (0.3 to 1.9 μg/mL) and major component precision of 0.4 to 2.1% RSD ($n = 6$) using a CAD [56].

The volatility of the dried aerosol particles was investigated by plotting response versus evaporator temperature. From Figure 1.20, particles with lower volatility exhibited increased response as this temperature was raised. However, there is a trade-off to this method of optimizing response since noise also increased with evaporator temperature. Here, the most volatile compounds exhibited the best sensitivity at a CNLSD evaporator temperature of 40°C, whereas sensitivity for less-volatile compounds was greatest at temperatures above 60°C.

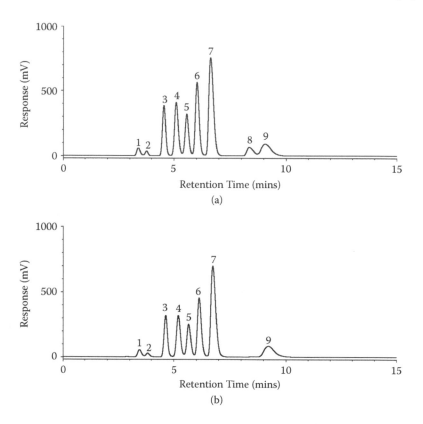

FIGURE 1.19 Separation of singly charged volatile bases on zic-*p*HILIC. Addition of 0.2 mM HCl to mobile phase enables detection of ammonia peak. Each analyte at 0.1 mg/mL using acetonitrile/water (1:1, v/v) diluent. 1, diisopropylethylamine; 2, triethylamine; 3, diethylamine; 4, isobutylamine; 5, morpholine; 6, ethylamine; 7, methylamine; 8, ammonia; 9, hydrazine. (a) Mobile phase = 0.2 mM HCl in trifluoroacetic acid/acetonitrile/water (0.04:60:40, v/v/v); (b) mobile phase = trifluoroacetic acid/acetonitrile/water (0.04:60:40, v/v/v), flow rate = 1.0 mL/min, injection volume = 5.0 μL, column temperature = 40°C. (Reprinted from R. D. Cohen, Y. Liu, and X. Y. Gong, *J. Chromatogr. A*, 1229: 172, 2012, with permission from Elsevier. Copyright 2012.)

1.6   RESPONSE UNIFORMITY

Aerosol-based detectors are often regarded as "universal," which is the opposite of being selective and refers to the ability to detect all sample components. Of course, this is not fully true since most volatile and some semivolatile molecules are difficult to detect. An additional property often tied to universality is that of response uniformity, whereby the detector can provide a consistent response irrespective of molecular structure. To a degree, this property is demonstrated in Figure 1.21 for the Corona CAD; a diverse set of 21 small-molecule pharmaceuticals, each at a concentration of 0.2 mg/mL, were eluted under reversed-phase, isocratic conditions. From this figure, several limitations are noteworthy: a lower response observed for semivolatiles and

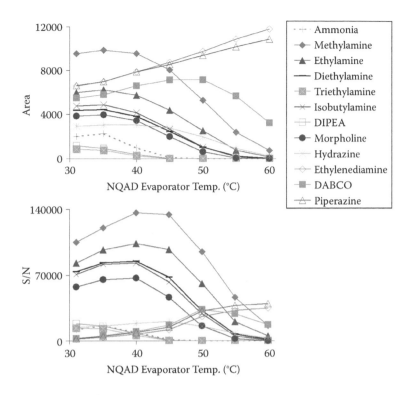

FIGURE 1.20 (See Color Insert.) Effect of CNLSD response versus evaporator temperature for volatile bases, each at a concentration of 0.1 mg/mL. Mobile phase = 0.2 mM HCl in trifluoroacetic acid/acetonitrile/water (0.04:60:40, v/v/v) used for monoprotic bases and mobile phase = trifluoroacetic acid/acetonitrile/water (0.4:70:30, v/v/v) used for polyprotic bases; column = zic-pHILIC; flow rate = 1.0 mL/min; injection volume = 5.0 µL; column temperature = 40°C. DIPEA = N,N-diisopropylethylamine (Reprinted from R. D. Cohen, Y. Liu, and X. Y. Gong, *J. Chromatogr. A*, 1229: 172, 2012, with permission from Elsevier. Copyright 2012.)

a higher response for bases, which were charged under the elution conditions. Both findings were previously discussed in Section 1.5.4, "Volatility Limit."

Nevertheless, the CAD provided a fairly consistent response independent of structure as long as the compounds were nonvolatile and had the same charge state under the elution conditions. From Table 1.4, the percentage RSD of the response factors was 4.8% for 0.2 mg/mL, a typical concentration for major component analysis. However, response uniformity decreased with concentration—dropping to 16.8% RSD when the set of 14 nonvolatiles was each injected at 0.2 µg/mL (or the 0.1% level, representative of typical impurities).

Response uniformity for the CAD has been found to be an improvement over the ELSD [19,79,82–84]. For the ELSD, response depends on additional properties of the RI and density of the dried aerosol particles, which are both dependent on molecular structure [13,33]. Although Koropchak and coworkers hoped that the CNLSD would

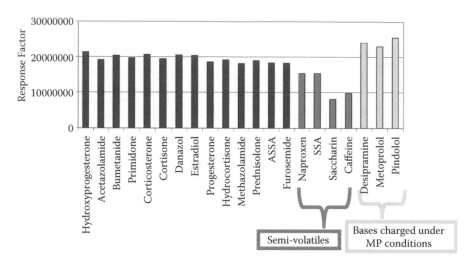

FIGURE 1.21 CAD uniformity of response. Concentration about 0.20 mg/mL; injection volume = 10 μL; mobile phase (MP) = trifluoroacetic acid/water/acetonitrile (0.1:35:65, v/v/v); column = xBridge C18; column temperature = 40°C; flow rate = 1.0 mL/min. SSA, salicylsalicylic acid; ASSA, acetylsalicylsalicylic acid.

provide more uniform response than the ELSD due to minimizing RI differences between particles, the opposite was found [44]. The reason is not known, but possible factors they cited were "(a) density differences which would influence the absolute sizes for the dry particle size distribution; (b) particle wettability (by butanol in this case) and particle morphology, which might influence the growth process; and (c) analyte specific factors which might influence particle coagulation or collection rates."

1.6.1 GRADIENT EFFECTS

All aerosol-based detectors exhibit a response proportional to the organic content of the mobile phase [50,75]. The reason for this behavior has primarily been attributed to improved nebulization efficiency due to reduced surface tension and viscosity of the eluent [13,75,85]. Table 1.5 lists physical properties for common reversed-phase solvents, showing much larger surface tension and viscosity for water. From the

TABLE 1.4

CAD Response Factor Uniformity for 14 Nonvolatile, Neutral Molecules; Conditions Same as in Figure 1.6

Analyte Concentration (mg/mL)	% RSD of Response Factors
0.2	4.8
0.002	10.3
0.0002	16.8

TABLE 1.5
Physical Properties for Common HPLC Solvents

Solvent	Surface Tension[a] (dyne/cm)	Viscosity[a] (cP)	Density[a] (g/mL)
Acetonitrile	19.10	0.38 (at 15°C)	0.7822
Methanol	22.55	0.59	0.7913
Tetrahydrofuran	26.4 (at 25°C)	0.55	0.888
Water	72.8	1.00	0.9982

Source: Data from http://macro.lsu.edu/HowTo/solvents.htm.

[a] Measured at 20°C unless otherwise indicated.

Nukiyama and Tanasawa equation [86], increased viscosity (μ in Pa•s) and surface tension (σ in N/m) result in larger mean droplet size (i.e., Sauter mean diameter) D_0:

$$D_0 = \frac{0.585\sqrt{\sigma}}{\left(v_g - v_l\right)\sqrt{\rho_m}} + 53.22\left[\frac{\mu}{(\sigma\rho_m)^{1/2}}\right]^{0.45}\left[\frac{Q_l}{Q_g}\right]^{1.5} \tag{1.7}$$

Other parameters are volumetric flow rate Q (m³/s), linear velocity v (m/s), and eluent density ρ_m (kg/m³); subscripts g and l refer to nebulizer gas and mobile phase, respectively. In the case of the Corona CAD, an impactor is employed to remove large aerosol droplets, which would dry inefficiently, causing higher background noise. At increased organic content, the wet aerosol particle distribution shifts away from *large* droplets per Equation (1.7). While this leads to smaller dried aerosol particles per Equation (1.1), the net result is an overall greater mass of analyte reaching the detector. Other detectors incorporate various strategies for large-droplet removal as previously discussed, thereby experiencing similar gradient effects.

Three different methodologies have been described to compensate for this effect during gradient elution. One involves constructing three-dimensional (3D) calibration curves, where the *x*-, *y*-, and *z*-axes are in dimensions of retention time, concentration, and response, respectively. In 2004, Mathews et al. applied this strategy to ELS detection using 5-fluorocytosine (5-FC) as a universal calibrant since it had desirable properties of being nonvolatile, nonretained, and soluble at all mobile phase compositions [26]. 5-FC, when dissolved in acetonitrile/water (8:2, v/v), also exhibited good peak shape on three columns [Luna C18, Synergi Hydro, and Luna C8(2)] under both low and high pH reversed-phase elution conditions using water and acetonitrile as weak and strong solvents, respectively. The 3D calibration curve was built over five runs using standard concentrations ranging from 0.037 to 3.0 mg/mL. For each run, the calibrant was injected eight times at regular intervals. Finally, the 3D response surface was built using interpolation software. This process is illustrated in Figure 1.22. To determine an unknown sample concentration, the equation was:

$$\log(Amount) = c_1 + c_2(R) + c_3[\log(A)] - c_4(R)^2 - c_5(R)[\log(A)] + c_6[\log(A)^2] \tag{1.8}$$

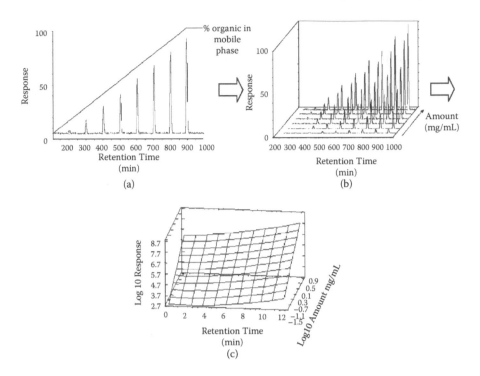

FIGURE 1.22 Steps used to construct 3D calibration curve. (A) Nonretained calibrant (5-fluorocytocine, 5-FC) is injected at regular intervals throughout gradient run. (B) Step A is repeated multiple times for different levels of calibrant. (C) The 3D response surface is constructed using interpolation software. (Reprinted with kind permission from Springer Science+Business Media from B. T. Mathews, P. D. Higginson, R. Lyons, J. C. Mitchell, N. W. Sach, M. J. Snowden, M. R. Taylor, and A. G. Wright, *Chromatographia*, 60: 625, 2004, Figure 3.)

where R is retention time, A is the sample's peak area, and c_1 through c_6 are constants obtained via interpolation. The accuracy of their method is demonstrated by the data in Table 1.6. The average error was 8.3%. The six steroids chosen all showed particularly high accuracy, with relative error less than 4%. Higher errors for outliers, in particular reserpine and 4-aminobenzophenone, may be due to either increased volatility or differences in RIs versus the calibrant.

Hutchinson and coworkers constructed a 3D calibration response curve using reversed-phase conditions for the CAD [79]. They chose four different calibrants to build the model: sucralose, amitriptyline, dibucaine, and quinine. Contrary to the recommendation of Mathews et al., these calibrants would be retained under some mobile phase conditions. Thus, the model was built without a column using FIA. The advantage of this approach was that it could be transferred to different columns by using a band-broadening correction factor obtained after only one run (note: all peaks experience similar band broadening when eluted by a linear solvent gradient).

TABLE 1.6
Accuracy of 3D Calibration Curve for ELSD

Compound	Prepared Concentration (mg/mL)	3D Estimated Concentration (mg/mL)	Relative Error (%)
Hydrocortisone	0.16	0.15	3.8
Cortisone	0.50	0.49	2.3
Cortisone	1.00	0.99	1.2
Cortisone	1.19	1.17	1.6
Adrenosterone	1.01	0.99	2.3
Dexamethasone	1.03	0.99	3.5
Prednisolone	1.00	0.97	2.8
9-Phenylanthracene	1.02	1.04	1.9
N-(3,5-Dinitrobenzoyl)-R-α-phenylglycine	1.02	0.80	21.3
Reserpine	1.01	0.73	28.2
Camphor p-tosylhydrazone	1.01	0.97	4.0
(Acetonyl-p-chlorobenzoyl)-4-hydroxycoumarin	1.02	0.89	12.8
Propranolol	0.50	0.55	9.6
Propranolol	1.00	1.12	11.7
Propranolol	1.09	1.21	11.2
4-Aminobenzophenone	1.07	1.30	21.7
Amitriptyline	1.25	1.26	0.7

Source: Table modified from B. T. Mathews, P. D. Higginson, R. Lyons, J. C. Mitchell, N. W. Sach, M. J. Snowden, M. R. Taylor, and A. G. Wright, *Chromatographia*, 60: 625 (2004).
Column = Luna C18; detector = Alltech 2000; evaporator = 40°C; gas flow = 3.0 L/min; MP flow = 0.3 mL/min; MP A = 0.1% formic acid in H_2O; MP B = acetonitrile; gradient = 5–95% B in 8 min, then 1-min hold.

A relative error of 12.8% was found when the model was tested using 12 compounds with diverse structures. However, three sets of outliers were noted and thus excluded from this result: Five volatile compounds had severely lower response, two quaternary ammonium compounds with permanent charges exhibited higher response, and two lipophilic compounds eluted outside the gradient window.

In a follow-up paper, Hutchinson et al. compared 3D response curves generated by FIA of sucralose (0.0001 to 1 mg/mL) on four different aerosol detectors (two CADs, an ELSD, and a CNLSD) shown in Figure 1.23 [50]. The two CADs (Corona CAD and Corona Ultra) produced plots with similar shapes, but the Corona Ultra plot had a maximum response nearly three times as high due to an improved nebulization design optimized for UHPLC analysis. The response for the ELSD (Varian 385-LC ELSD) topped out at high concentrations due to limited dynamic range (which is typical of many older-model ELSDs), and it exhibited a characteristic sigmoidal shape

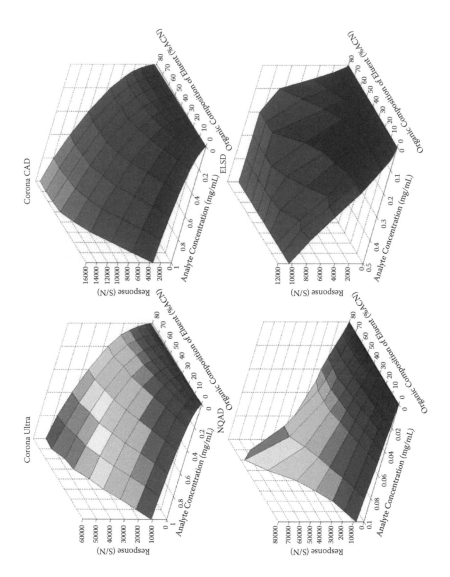

FIGURE 1.23 (See Color Insert.) Gradient response curves for four aerosol-based detectors. ACN = acetonitrile. (Reprinted from J. P. Hutchinson, J. F. Li, W. Farrell, E. Groeber, R. Szucs, G. Dicinoski, and P. R. Haddad, *J. Chromatogr. A*, 1218: 1646, 2011, with permission from Elsevier. Copyright 2011.)

at nearly all organic compositions. The CNLSD (Quant NQAD QT-500) achieved the highest sensitivity of all the evaluated aerosol detectors at 60% organic content. Also noteworthy, deviations from linearity for this detector appeared to be more pronounced at higher organic content.

Recently, Hutchinson and coworkers looked at the impact different HILIC conditions (viz., different strong solvents such as acetone, ethanol, isopropanol, and methanol) had on the CAD's 3D calibration curve for a series of carbohydrates [69]. To generate the calibration curve, they used quinine as a calibrant since sucralose, which was used in the previous two studies, was insoluble in ethanol and isopropanol. The change in response with respect to organic content was more linear for ethanol, isopropanol, methanol, and acetone compared to acetonitrile, thereby making the gradient correction simpler.

The second methodology, referred to as inverse gradient compensation, was put forth by Gorecki and coworkers [87,88]. It involves using a separate pump to deliver the exactly opposite gradient composition relative to the chromatographic separation. The two liquid streams are mixed using a tee just before the aerosol detector, with the net result being that the detector experiences a constant mobile phase composition throughout the run. This is demonstrated in Figure 1.24. Differences in calibration curves obtained by this method for various organic contents did not exceed 5% [88]. The obvious advantage of this approach over the previous methodology is that it is independent of a specific calibrant. The downsides are the expense of an additional pump and difficulty with obtaining identical dwell volumes up to the mixing tee. This task can be made somewhat easier by using identical columns on each gradient line, such that the column experiencing the inverse gradient is only needed to ensure the correct dwell time.

A third methodology, demonstrated in Figure 1.25, involves varying the *evaporator* gas flow throughout the run. This feature, referred to as real-time gas control, is found on Agilent ELSDs. Mourey and Oppenheimer showed that varying *nebulizer* gas flow had the greatest impact on response compared to other instrument parameters [13]. Specifically, lower flow rates resulted in increased response due to a shift to larger particle distributions with improved scattering efficiencies. A similar change in response can be obtained by adjusting the evaporator gas flow rate (see Figure 1.26).

From Figure 1.25, a low evaporator gas flow rate was used at the start of the run when the aqueous component was 95%, and this flow rate was nearly tripled by the end of the gradient when the aqueous component was only 5%. This resulted in a marked improvement in response uniformity throughout the run. Since this feature was recently added, we are not aware of any studies that independently evaluated it versus other gradient compensation methods.

1.6.2 UNIVERSAL CALIBRATION

The purpose of this section is to look at the application of single-calibrant strategies to quantify sample concentrations without an isolated standard for each particular compound. Due to high uniformity of response as previously discussed, there has been great interest in this goal for aerosol-based detectors, even dating back to

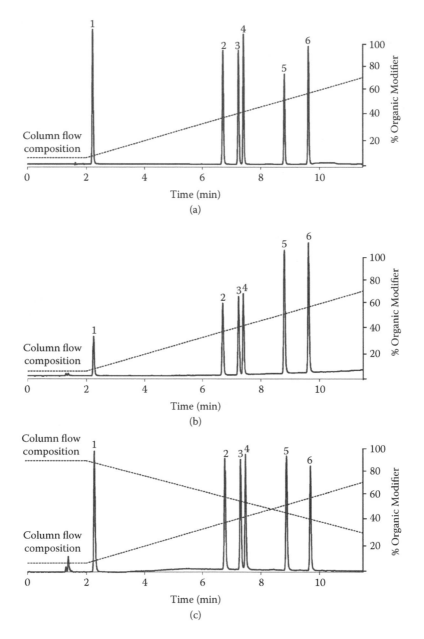

FIGURE 1.24 Chromatograms demonstrating utility of gradient compensation for sulfonamides of nearly equal concentration (100 mg/L). (a) UV; (b) CAD without gradient compensation; (c) CAD with gradient compensation. 1, Sulfaguanidine; 2, sulfamerazine; 3, sulfamethazine; 4, sulfamethizole; 5, sulfamethoxazole; 6, sulfadimethoxin. (Reprinted [adapted] with permission from T. Gorecki, F. Lynen, R. Szucs, and P. Sandra, *Anal. Chem.*, 78: 3186, 2006. Copyright 2006 American Chemical Society.)

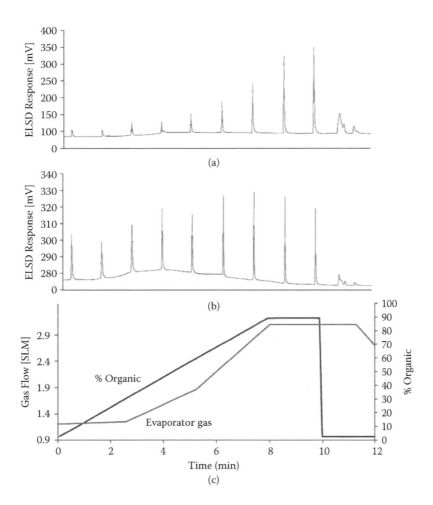

FIGURE 1.25 5-Fluorocytocine (5-FC) injected without a column at 1-min intervals across gradient from 5 to 95% acetonitrile. ELSD gradient effects (a) were overcome by programmable evaporator gas flow gradient resulting in uniform response (b). Gas flow measured in units of standard liters per minute (SLM). (Adapted from *Agilent Technical Overview*, Agilent Technologies, Inc., Santa Clara, CA, April 1, 2012, http://www.agilent.com/chem/elsd, 5990-9159EN)

Charlesworth's publication in 1978 [12]. The value is considerable, for example, time savings from not isolating/preparing individual reference standards especially for labile compounds, more accurate purity determinations, better characterization of complex mixtures, and higher-throughput analyses since only a single set of standards has to be prepared and injected.

Often, aerosol-based detectors have been described as mass detectors, meaning that response is dependent on the mass of analyte present as opposed to the concentration in solution. This is a primary reason that the inverse gradient compensation technique previously discussed resulted in negligible sensitivity change for the

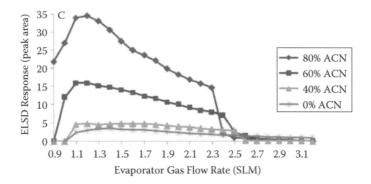

FIGURE 1.26 Effect of ELSD evaporator gas flow rate on detector response for different levels of organic content. Gas flow measured in units of standard liters per minute (SLM). (Reprinted from J. P. Hutchinson, J. F. Li, W. Farrell, E. Groeber, R. Szucs, G. Dicinoski, and P. R. Haddad, *J. Chromatogr. A*, 1218: 1646, 2011, with permission from Elsevier. Copyright 2011.)

CAD even though concentration was diluted by a factor of 2 [87]. Other universal LC detectors, such as low-wavelength UV and MS with current atmospheric pressure ionization sources, have responses that depend highly on molecular properties. Low-wavelength UV response depends on molar absorptivity, and there can be several orders of magnitude difference between structurally dissimilar compounds. For MS, response is dependent on ionization efficiency, which also can be orders of magnitude different between compounds.

Mojsiewicz-Pienkowska [89] reviewed universal calibration applied to the ELSD. Difficulties with universal calibration, particularly for this detector, are nonlinear response curves (sigmoidal over large ranges); moderate limits of detection; in some cases poor dynamic range for certain ELSD models; deviations from uniformity, particularly for volatiles and semivolatiles; and gradient effects. Investigators have tried various approaches to compensate for these deficiencies, such as linearizing the response curve, attempting to correlate volatility with predictions of physical properties, and adjusting for gradient effects via several methodologies.

Fang et al. evaluated a single-calibrant approach with ELSD to rapidly determine the purity of compounds in a drug discovery combinatorial library [90]. Such libraries can contain many thousands to millions of compounds, making it increasingly difficult to rapidly assess their purity both when first synthesized and over time. This problem is further complicated by factors that include unknown stability and limited sample quantities. Since compound purity is directly tied to the potency of biological assays, this can be a major issue affecting both lead identification and optimization. In their study, Fang and coworkers built one calibration curve from six standards using Equation (1.5). They then applied this calibration curve to determine the purity of 84 other structurally diverse compounds of known quality and reported an average quantitation error of 18.5%. Other studies looking at single-calibrant ELSD with smaller sample sets reported errors from 10 to 20% using structurally similar standards [91,92]. Fries et al. used both isocratic and gradient reversed-phase HPLC with an ELSD to determine the level of metabolites with a single standard, but they found inadequate sensitivity for routine use [93]. It should be pointed out that none of these studies employed a correction for gradient effects or directly accounted for low-volatility compounds.

A comprehensive investigation on the accuracy of ELSD for single-calibrant quantitation was conducted by researchers from GlaxoSmithKline [77]. They found ELSD to be less accurate than the chemiluminescent nitrogen detector (CLND). Quantitative ^{1}H-NMR using the electronic reference to access in vivo concentrations (ERETIC) method was considered a "gold standard" for providing absolute purity, although this technique does not lend itself to high-throughput analysis [94]. The accuracy of HPLC with either ELSD or CLND compared to NMR for purity determinations of 117 samples is presented in Figure 1.27. A perfectly accurate detector would have all data points on the line $y = x$. Clearly, the CLND exhibited less deviation than the ELSD. Furthermore, the ELSD data were biased toward reporting lower concentrations.

Squibb et al. attributed their bias due to lower response for semi-volatiles and to mismatch between the mobile phase with the calibrant and sample diluents [81]. In place of a single, non-retained calibrant, they chose four calibrants spanning the range of the HPLC retention time window. The calibrants were dissolved in the same diluent as the samples. For a test set of 28 non-volatile compounds, good agreement was found relative to the known concentrations (mean accuracy from 94%–99% for replicate determinations over three days). Their method's accuracy ranged from 70–132% for concentrations measured between 0.72 to 2.56 mg/mL. Two compounds, acetanilide and ibuprofen, gave remarkably lower results due to volatility, and these were excluded.

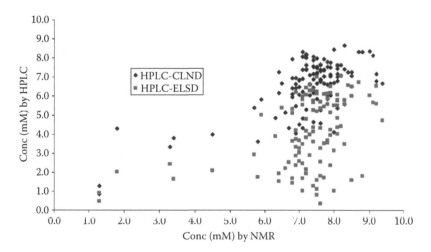

FIGURE 1.27 Comparison of chemiluminescent nitrogen detection (CLND) and ELSD for single-calibrant determination of 117 samples. "Exact" results from quantitative nuclear magnetic resonance (NMR). (Reprinted [adapted] with permission from S. Lane, B. Boughtflower, I. Mutton, C. Paterson, D. Farrant, N. Taylor, Z. Blaxill, C. Carmody, and P. Borman, *Anal. Chem.*, 77: 4354, 2005. Copyright 2005 American Chemical Society.)

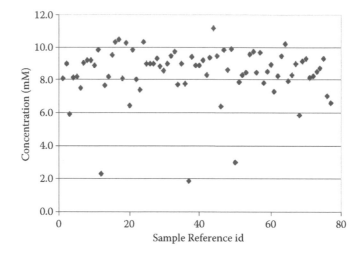

FIGURE 1.28 Accuracy of single-calibrant determination using CAD for 77 known samples dissolved at 10 mM in DMSO. (Reprinted by permission of SAGE Publications from I. Sinclair and I. Charles, *J. Biomol. Screen.*, 14: 531, 2009. Copyright © 2009 by Society for Laboratory Automation and Screening.)

Several investigators applied the single-calibrant methodology by using a CAD, hopeful that results would be better than past studies with an ELSD. Lämmerhofer and coworkers looked at ionic liquids and found high accuracy (90–108%) for nine compounds [95]. However, they did find two outliers, benzoic acid and thiosalicylic acid, exhibited very poor response, presumably due to higher volatility. Sinclair and Charles also used the CAD to determine the accuracy of preparing known solutions of 77 compounds in dimethylsulfoxide (DMSO) at a level of 10 mM [96]. Their results are shown in Figure 1.28. The mean concentration was 8.4 mM with a standard deviation of 1.6. Several low-concentration outliers again may have been due to volatility, although the authors believed low DMSO solubility was the culprit.

1.7 APPLICATIONS

In the next four sections, the application of aerosol-based detection to several important problems where analytes lack a UV chromophore is discussed. The applications chosen were the analysis of counterions, lipids, carbohydrates, and formulation excipients. Several other problems for which aerosol-based detectors have often been used are in the analysis of amino acids [97], peptides [98], polymers [99,100], pharmaceuticals (especially aminoglycoside antibiotics) [54,59,101–103], Chinese herbal medicines [53,62], and cleaning validation studies [104].

1.7.1 APPLICATION 1: COUNTERION ANALYSIS

Active pharmaceutical ingredients (APIs) are often produced as salts, using pharmaceutically acceptable counterions, to improve bioavailability and physicochemical stability. Maintaining and accurately monitoring the ratio of counterion to active ingredient is necessary to ensure API integrity. Various separation and detection methods have been used for this type of analysis. Among them, ion chromatography with conductivity detection has been widely used [105–108]. However, this methodology suffers from several disadvantages, such as costly and specialized equipment, poor recoveries when hydrophobic compounds are present due to electrode fouling, and separate conditions needed to analyze cations and anions [109,110].

In 1997, Lantz et al. first investigated the use of an ELSD to measure the levels of chloride counterion, drug substance, and process-related impurities in a single HPLC run using mixed-mode columns composed of either C18/anion, phenyl/anion, or C8/cation bonded phases [111]. Subsequently, this work was followed up by a study in 2003 in which the level of piperazine counterion was determined under HILIC conditions on a cyano column [112]. In 2005, Pack and Risley separated sodium, lithium, and potassium counterions by HILIC using a monolithic silica column, and they found a relative error for sodium of 3% with respect to the theoretical composition for three different APIs [113]. Finally, in 2006, Risley and Pack reported on the simultaneous separation of acidic and basic counterions by a zwitterionic sulfobetaine stationary phase (zic-HILIC) [114]. While the separation mechanism on this stationary phase was complex due to the interdependence of buffer strength, pH, and organic content, it afforded more parameters to manipulate to obtain baseline resolution.

The mechanism by which ions are separated under HILIC conditions, and zwitterionic stationary phases in particular, has been studied by several investigators, but details are still being debated [115]. Generally, it is believed that polar ions partition into an adsorbed water layer on the stationary phase's surface while experiencing electrostatic effects. Because of the complexity of the separation mechanism, Dejaegher et al. recommended a multivariate approach to HILIC method development [116]. Nevertheless, Risley and Pack have identified several guidelines for method development using a zic-HILIC column [114]:

1. Cation retention decreases and anion retention increases with buffer strength.
2. Cation retention increases and anion retention decreases with buffer pH.
3. Retention decreases for both cations and anions by increasing aqueous content.

Recent studies of counterion analysis have utilized the CAD for improved precision, sensitivity, dynamic range, and linearity over the ELSD [70,73,82,117,118]. In particular, Crafts et al. developed and validated a gradient method using the zic-*p*HILIC column to determine nitrate, chloride, phosphate, sulfate, potassium, and sodium counterions [117]. To improve selectivity, methanol, isopropanol, and acetonitrile were used as weak solvents. Elution of divalent anions was found to be particularly sensitive to methanol, while isopropanol exhibited a slightly less-dramatic effect. Baseline noise was minimized by using the same buffer concentration in mobile phases A and B. Acceptable linearity ($r^2 \geq 0.933$), accuracy (percentage recoveries of 95–105%), and precision (%RSD < 5%) were found over the range from 0.8 to 100 µg/mL. Detection limits were 1–7 ng. Dynamic range was sufficient to

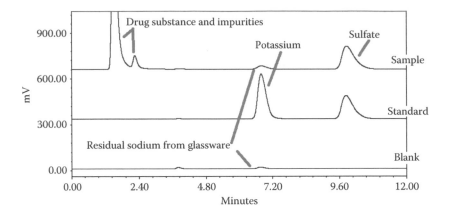

FIGURE 1.29 Sulfate counterion determination using HILIC conditions. Detection = CAD; standard = K_2SO_4; mobile phase = 50 mM ammonium acetate, pH 4.8/acetonitrile (33:67, v/v); column = zic-*p*HILIC (150 × 4.6 mm, 5 μm); flow rate = 1.0 mL/min; column temperature = 40°C; injection volume = 5 μL.

quantitate both a 0.1% w/w sodium impurity, as well as a 0.7 mg/mL drug substance peak in the same run, which the authors claimed unobtainable by current ELSDs at the time.

An example chromatogram from a CAD showing both the drug substance and sulfate counterion is shown in Figure 1.29. There are two remarkable points about this chromatogram. First, since the elution was done under HILIC conditions, most drug substances and impurities would elute near the void, making method development significantly easier. The suitability of HILIC can be quickly assessed by a compound's lipophilicity (note: this can be estimated quickly via ChemDraw). If a compound's log *P* (octanol-water partition coefficient) or log *D*, for ionizable compounds, is greater than 0, then it will generally be unretained. Second, notice that sodium is present in both the blank, standard, and sample solutions. Sodium is a common contaminant of borosilicate glassware. By using plasticware and LC-MS-grade water, sodium contamination can be minimized.

Huang et al. quantified counterions by HPLC-CAD using the zic-*p*HILIC column, which is a polymeric version of zic-HILIC, due to improved baseline noise at neutral pH [73]. Two isocratic methods were developed for either monovalent (nitrate, chloride, bromide, sodium, potassium) or divalent (calcium, magnesium, sulfate, phosphate) ions. Performance characteristics were then compared versus IC (see Table 1.7), and they noted several distinct advantages: "Compared to IC with conductivity detection, HILIC/CAD technique was easier and less expensive to implement, required much less time for method development, and provided adequate precision and overall better accuracy [73]."

In a most recent study, Zhang et al. described the development of an HPLC-CAD method for the determination of 25 commonly used pharmaceutical counterions using a mixed-mode column [70]. This column, Acclaim Trinity P1, contains porous silica particles coated with charged nanopolymer beads, and it displays

TABLE 1.7
Accuracy Data for Counterion Analysis in Various Drug Substances Comparing HPLC-CAD to IC

| Compound | Counterion | Theoretical Weight Percentage | HPLC-CAD Results | | IC Results |
			Percentage of Theory	Percentage RSD ($n = 3$)	Percentage of Theory
A	HCl	9.0	98.0	0.52	97.8
B	HCl	5.4	99.3	0.32	94.8
C	HCl	7.3	98.9	0.54	97.3
D	HCl	7.7	100.0	0.68	Not determined
E	K^+	5.3	101.6	0.68	98.5
F	Ca^{2+}	3.8	98.7	0.89	98.1

Source: Modified from Z. Huang, M. A. Richards, Y. Zha, R. Francis, R. Lozano, and J. Ruan, *J. Pharm. Biomed. Anal.*, 50: 809 (2009).

reversed-phase as well as weak anion and strong cation exchange properties [119]. The effects of column temperature and various mobile phase parameters (viz., organic content and buffer salt type, pH, and concentration) were studied, and an optimized gradient separation was developed that resolved all 25 counterions within 20 min (see Figure 1.30).

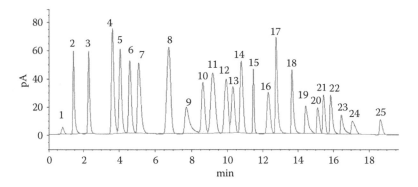

FIGURE 1.30 Separation of 25 common pharmaceutical counterions on Acclaim Trinity P1 column. Detector = Corona Plus CAD; buffer = pH 4.0 ammonium formate 200 mM; column temperature = 35°C; 1, lactate; 2, procaine; 3, choline; 4, tromethamine; 5, sodium; 6, potassium; 7, meglumine; 8, mesylate; 9, gluconate; 10, maleate; 11, nitrate; 12, chloride; 13, bromide; 14, besylate; 15, succinate; 16, tosylate; 17, phosphate; 18, malate; 19, zinc; 20, magnesium; 21, fumarate; 22, tartrate; 23, citrate; 24, calcium; 25, sulfate. (Reprinted from K. Zhang, L. L. Dai, and N. P. Chetwyn, *J. Chromatogr. A*, 1217: 5776, 2010, with permission from Elsevier. Copyright 2010.)

1.7.2   APPLICATION 2: LIPID ANALYSIS

Lipids are small molecules with diverse structures known for their hydrophobic or amphiphilic properties. They have been highly studied because of their biological roles in signaling and energy storage and as the primary structural components of membranes. There are several types, including glycerols, fatty acids, phospholipids, sterols, and carotenoids.

Many important lipid classes exhibit poor UV sensitivity and low volatility, making aerosol-based detection an obvious choice. The first publications on lipid analysis by HPLC with ELSD appeared in the early to mid-1980s [22,120]. Today, this approach is one of the most popular due to adequate sensitivity and compatibility with gradient elution and a broad range of nonpolar solvents. Normal-phase or nonaqueous reversed-phase elution conditions are frequently used because of the high *log P* (octanol-water partition coefficient) of most lipids [121]. While the normal phase has been better at separating lipids of different classes, the reversed phase has been applied to lipids within the same class [122].

Recently, the CAD has been increasingly used for lipid analysis due to improved performance versus the ELSD [51,123,124]. Moreau studied the relationship between lipid structures and charged aerosol detectability [125]. Common acylglycerols (triglycerides, phospholipids, and glycolipids) and sterols were readily detected, while most fatty acid methyl esters were too volatile and therefore not detected. Semivolatile free fatty acids exhibited diminished response, but reproducible quantitative results were still achievable.

Merle et al. compared the ELSD and CAD for the separation of stratum corneum lipids (viz., cholesterol, fatty acids, and ceramides) by normal-phase HPLC [124]. Using an eluent of heptane, chloroform, and acetone with a polyvinyl alcohol functionalized silica (PVA-Sil) column, several different ceramides were separated from fatty acids and cholesterol. Notably, diluent polarity had a significant impact on resolution between C22:0 fatty acids with cholesterol [126]. Calibration via external standards for each lipid class was found to be superior to a single internal standard since gradient effects were not corrected. The CAD demonstrated superior precision and accuracy over the ELSD; therefore, it was selected for the final method conditions. An average concentration ratio from forearm extracts of 20:20:60 for cholesterol, fatty acids, and ceramides, respectively, agreed with past stratum corneum studies.

Lísa et al. developed a nonaqueous reversed-phase method using inverse gradient compensation for a CAD to determine the triacylglycerol composition of plant oils without external standard calibration [68]. Results for seven plant extracts were in good agreement with those obtained using APCI LC-MS where individual external standards had to be applied. The response uniformity comparing APCI and CAD is shown in Table 1.8 for triacylglycerols ranging in carbon chain length from 7 to 22. APCI response factors were notably much higher for low molecular weight compounds due to "strong discrimination of low *m/z* values in the ion trap analyzer." CAD response factors were more uniform, although there was a dependency on chain length, particularly noticeable at the extremes. The reason for this behavior was not known. Nevertheless, this had little impact on the accuracy of determining triacylglycerol composition of most plant oils since carbon chain length typically ranged from 16 to 18 with only a few exceptions.

TABLE 1.8
Relative Response Factors (RRFs[a]) of 19 Single-Acid Triacylglycerol Standards Comparing CAD with Inverse Gradient Compensation versus APCI-LC-MS

Compound (No. of Carbons: No. of Double Bonds)	Retention Time (min)	CAD RRFs	APCI[b] RRFs
C7:0	8.3	0.54	97.20
C8:0	10.1	0.74	74.44
C9:0	12.9	0.83	38.91
C10:0	16.7	0.86	17.62
C11:0	21.6	0.89	10.85
C12:0	27.4	0.94	6.04
C13:0	33.8	0.97	4.31
C14:0	40.1	0.95	2.77
C15:0	46.0	0.98	1.75
C16:0	51.1	1.01	1.32
C17:0	55.6	1.01	0.81
C18:3	24.1	0.92	0.40
C18:2	35.8	0.98	0.57
C18:1	**48.6**	**1.00**	**1.00**
C18:0	59.3	1.02	0.61
C19:0	62.8	1.05	0.49
C20:0	65.5	1.11	0.40
C21:0	68.2	1.27	0.39
C22:0	70.5	1.38	0.46

Source: Modified from M. Lísa, F. Lynen, M. Holcapek, and P. Sandra, *J. Chromatogr. A*, 1176: 135 (2007).

[a] Response factors were calculated with respect to triolein (C18:1), highlighted in bold above.

[b] APCI data from M. Holcapek, M. Lisa, P. Jandera, and N. Kabatova, *J. Sep. Sci.*, 28: 1315 (2005).

Schönherr et al. developed and validated a reversed-phase HPLC-CAD method to analyze the components of a liposomal formulation, a recently popular approach to drug delivery [67]. They demonstrated acceptable repeatability (RSD from 0.5 to 3.7%), recovery (98 to 103%), detection limits (0.2–0.4 µg/mL), and linearity for one order of magnitude. An example chromatogram from their formulation is shown in Figure 1.31. They found detection by a CAD to be more suitable than low-wavelength UV due to an unacceptably high rising baseline from their methanol gradient.

Chaminade and coworkers analyzed pegylated phospholipids in microcapsule suspensions of perfluorooctyl bromide, an ultrasound contrast agent [128]. They developed a simple 20-min nonaqueous reversed-phase method using a CAD to directly quantify pegylated phospholipids both free in solution and associated with microcapsules. A power function provided the best fit to the calibration data over the range from 2.23 to 21.36 µg/mL. Acceptable accuracy (90 to 115%), precision (RSD ≤ 5.3%), quantitation limits (100 ng), and selectivity were obtained.

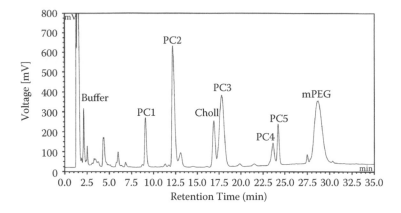

FIGURE 1.31 Reversed-phase HPLC-CAD separation of components from liposomal formulation. Column = xBridge C18 (3.5 μm, 150 × 3 mm); mobile phase A = trifluoroacetic acid/acetonitrile/water (0.05:90:10, v/v/v); mobile phase B = 0.05% v/v trifluoroacetic acid in methanol; injection volume = 10 μL; flow rate = 0.5 mL/min; gradient = 40 to 100% B in 25 min followed by 10 min hold; column temperature = 35°C; detection = Corona Plus. PC1–PC5, phosphatidylcholine; Chol, cholesterol; buffer, phosphate; mPEG, *N*-carbonyl-methoxy(polyethylene glycol 2000)-1,2-distearoyl-sn-glycero-3-phosphoethanolamine. (Reprinted from C. Schönherr, S. Touchene, G. Wilser, R. Peschka-Suss, and G. Francese, *J. Chromatogr. A*, 1216: 781, 2009, with permission from Elsevier. Copyright 2009.)

1.7.3 APPLICATION 3: ANALYSIS OF FORMULATION EXCIPIENTS

Excipients are loosely defined as pharmacologically inactive ingredients possessing a multitude of uses in drug formulations. For example, they are commonly used as fillers, disintegrants, binders, coatings, flavorants, lubricants, preservatives, glidants, and sorbents. A considerable amount of recent formulation research has been directed toward the design and application of excipients to enhance bioavailability and physicochemical stability of APIs [129–132]. This trend has largely been driven by an increase in drug development candidates with low solubility and poor permeability. In this section, particular attention is paid to the application of aerosol-based detectors for the analysis of these compounds.

Polysorbates are amphiphilic, nonionic surfactants derived from the fatty acid ester of polyoxyethylene sorbitan. The most common polysorbates used as excipients are polysorbate 20 (polyoxyethylenesorbitan monolaurate) and polysorbate 80 (polyoxyethylenesorbitan monooleate). They have been used in formulations not only as delivery vehicles for both small molecules and protein-based drugs but also to retain the biological activity of proteins on storage. Since polysorbates have weak chromophores, previous methods employed low-wavelength UV detection or derivatization. These methods were time consuming and had poor sensitivity [133,134].

Fekete et al. analyzed polysorbate 80 in the presence of a four-helix bundle protein by using HPLC coupled with a CAD [135]. A gradient method was developed using a poroshell 300SB-C18 column and acetonitrile–methanol–water–triflouroacetic acid as the mobile phase. Specificity was demonstrated by baseline resolution of polysorbate 80 from native, oxidized, and reduced proteins. Method robustness was demonstrated, and the limit of quantitation was 10 µg/mL.

In another study, Fekete et al. examined polysorbate 20 and PEGs using the CNLSD [64]. They developed a fast-gradient HPLC method using a core shell kinetex C18 column with methanol, water, and TFA as the mobile phase. The method gave baseline resolution for the major components, polysorbate, PEG, and protein. Notably, the shape of the calibration curve was convex for concentrations from 20 to 140 µg/mL, and it was concave at higher concentrations. However, a linear fit was acceptable over a narrower range from 20 to 60 µg/mL. Also, the calibration curve was more linear at lower evaporator temperatures. Precision less than 6.0% was demonstrated by multiple injections over several days, and the quantitation limit of 10 µg/mL was established.

PEGs are synthetic, water-soluble polymers widely used as emulsifier surfactants in cosmetics, food, and pharmaceutical products. They can also be covalently attached to APIs (typically proteins or peptides) to make pegylated pharmaceuticals. Harris and Chess stated that pegylation "reduces immunogenicity, improves the solubility and bioavailability of medicinal proteins, reduces susceptibility to degradation by proteases, and decreases toxicity" [136].

Pharmaceutical analysis of PEG requires determination of both purity and polydispersity, which is a measure a polymer's molecular weight distribution. Kou et al. studied a 32-kDa PEG reagent using SEC with either a CAD or an ELSD [84]. The SEC method employed two serially coupled PLgel MIXED-D columns with dimethyl formamide as both eluent and diluent. The CAD response ($r = 0.9957$) was more linear than the ELSD's ($r = 0.9856$). In addition, the y-intercept for the CAD calibration curve was positive, while the ELSD's was negative and quite large. Consequently, when area percentage quantitation was performed, low-level impurities were overestimated by the CAD, while they were severely underestimated by the ELSD. As an example, the authors showed that dimer impurity levels from the ELSD were 5–36% of the corresponding CAD results. As dimer concentration decreased, these differences were magnified. Furthermore, the ELSD gave a disproportionately lower response to low-concentration fractions of PEG, which can cause an underestimation of polydispersity as shown in Figure 1.32. Polydispersity results closer to unity signify a narrower distribution of polymer molecular weights. Kou et al. found the CAD more accurately and better differentiated polydispersity; therefore, it was a more appropriate analytical tool than the ELSD for quality control.

Takahashi et al. also compared the CAD with the ELSD. They used CRM PEG 1000, an equal-mass mixture of PEGs, as analytes [99]. Ideally, peak areas would be the same regardless of the degree of polymerization n since the PEG used was a uniform oligomer. CAD response was uniform with respect to n at concentrations ranging from 0.4 to 10 mg/mL. For the ELSD, similar response was only observed for oligomers with n from 8 to 25 at the 10-mg/mL concentration. At lower concentrations (0.4 and 1 mg/mL), response varied from $n = 8$ to 25. In addition, peak intensities decreased for both detectors with increasing degrees of polymerization. However, the deviation at larger n was more pronounced for the ELSD. This difference had important implications with respect to evaluation of molecular mass distribution for polymers such as PEG. The molecular mass distribution is a parameter in

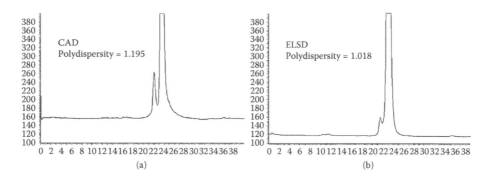

FIGURE 1.32 Comparison of PEG polydispersity for same lot obtained via (a) CAD and (b) ELSD. The minor peak is a dimeric impurity. Column = two coupled PLgel MIXED-D (5 μm, 300 × 7.5 mm); column temperature = 30°C; flow rate = 0.5 mL/min; mobile phase = dimethyl formamide. (Reprinted from D. W. Kou, G. Manius, S. D. Zhan, and H. P. Chokshi, *J. Chromatogr. A*, 1216: 5424, 2009, with permission from Elsevier. Copyright 2009.)

polymer chemistry that describes the relationship between the number of moles of each polymer species and the molar mass of that species. Per Figure 1.33, the authors showed that the ELSD greatly overestimated the fractions of PEG in the region $n = 18$ to 26. On the other hand, the CAD exhibited close agreement with the certified values. This was reflected in the differences from the certified value of weight; for example, the average molecular mass was 4.6% for the ELSD, whereas for the CAD, it was only 2.9%.

Takahashi et al. also compared the CNLSD to the ELSD using an equal-mass mixture of PEG as analyte [100]. Similar to the CAD, the CNLSD exhibited higher, more

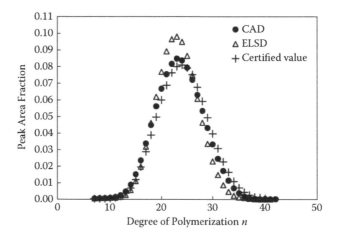

FIGURE 1.33 Comparison of mass fractions for CRM PEG 1000 obtained by a CAD and an ELSD versus certified values. (Reprinted from K. Takahashi, S. Kinugasa, M. Senda, K. Kimizuka, K. Fukushima, T. Matsumoto, Y. Shibata, and J. Christensen, *J. Chromatogr. A*, 1193: 151, 2008, with permission from Elsevier. Copyright 2008.)

uniform response at low-PEG concentrations. However, both the ELSD and CNLSD were saturated at high concentrations. The CNLSD was shown to be accurate for molecular mass distribution measurements by comparison against the certified values.

1.7.4 APPLICATION 4: CARBOHYDRATE ANALYSIS

Carbohydrates (also known as saccharides or sugars) are organic compounds composed of carbon, hydrogen, and oxygen, most commonly having an empirical formula of $C_m(H_2O)_n$. They have many important biological functions, such as for energy storage (e.g., starch and glycogen), for cell structure (e.g., cellulose and chitin), and as DNA and RNA components.

Carbohydrates are highly polar and generally exhibit poor UV absorption, making analysis less feasible by reversed-phase HPLC with UV detection [137]. Anion exchange chromatography with pulsed amperometric detection has frequently been performed; however, there are some drawbacks, such as low column capacity and incompatibility with MS [138–140]. Antonio et al. developed an electrospray ionization (ESI) MS method using a porous graphitic carbon (PGC) stationary phase for sugars and sugar phosphates [141]. Carbohydrate separations under HILIC conditions were studied as long ago as 1975 by Linden and Lawhead [142]. They used bare silica and amino stationary phases with a differential refractometer. In recent years, HILIC separations with aerosol-based detection have become more common. This has been driven by improvements in both column chemistries and detector sensitivity.

Asa described the analysis of monosaccharides and oligosaccharides (up to seven units) by isocratic HILIC elution with a CAD [143]. Baseline resolution was achieved for lactose, sucrose, glucose, and fructose. The method was reported to have detection limits in the low-nanogram range and was highly reproducible (RSDs < 2.0%).

Mitchell and coworkers compared HILIC and reversed-phase separations with ELSD, CAD, and ESI-MS for 12 polar molecules, including 4 saccharides [144]. Better sensitivity was achieved for all detectors by HILIC using a cross-linked diol column than reversed-phase elution.

Hutchinson et al. studied the effect of organic modifier on the separation and detection of 11 sugars (mono-, di-, and trisaccharides) with the CAD [69]. A polymeric amino column, asahipak NH2P-50 4E (100×3.2 mm, 5 µm), was used. This column is known for less bleed and improved stability over silica-based amino phases. They evaluated acetonitrile, acetone, ethanol, isopropanol, and methanol as weak eluents. Acetonitrile provided the highest efficiency and lowest background noise. Good response was observed for all analytes except ribose, which the authors attributed to Schiff base formation with the stationary phase. Gradient methods with acetonitrile or acetone as organic modifier were used to analyze a lager beer sample as shown in Figure 1.34. Acetone, a cheaper and greener alternative to acetonitrile, exhibited only moderately reduced performance.

Godin and coworkers used a classical anion exchange column (CARBOSep CHO-682 Pb) with a CAD to analyze lignocellulosic biomass [145]. A neutral detergent extraction (NDE), followed by sulfuric acid hydrolysis (SAH) of the structural polysaccharides was performed. The method was used to quantify cellulose, hemicellulose, xylan, arabinan, mannan, and galactan with acceptable precision and selectivity.

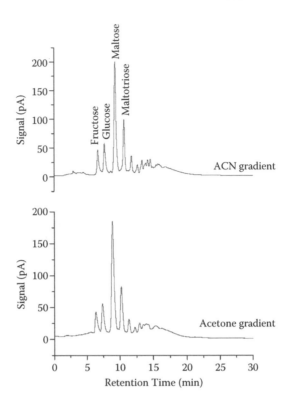

FIGURE 1.34 Comparison of acetone and acetonitrile as weak eluent for gradient HILIC separation of carbohydrates from a lager beer sample during fermentation. Detection = CAD; column = asahipak NH2P-50 4E (250 × 4.6 mm, 5 μm); column temperature = 30°C; mobile phase A = acetonitrile or acetone; mobile phase B = water; gradient = 90 to 10% A in 30 min for acetonitrile or 80 to 10% A in 30 min for acetone; flow rate = 1 mL/min; injection volume = 5 μL. (Reprinted from J. P. Hutchinson, T. Remenyi, P. Nesterenko, W. Farrell, E. Groeber, R. Szucs, G. Dicinoski, and P. R. Haddad, *Anal. Chim. Acta*, 750: 199, 2012, with permission from Elsevier. Copyright 2012.)

For longer-chain glycoproteins, traditional methods use chemical or enzymatic cleavage, then derivatization with a fluorescent label; finally, the oligosaccharides are analyzed by HPLC-FLD (fluorescent light detection). There are several drawbacks to this approach; for example, it is time consuming and often gives poor recoveries [146–148]. Inagaki et al. studied sialylglycopeptide (SGP), an oligosaccharide, containing 11 sugar units [66]. Two separation methods were developed; one was reversed-phase HPLC with derivatized SGP (Fmoc-SGP), and the other was a direct analysis by HILIC with an amino column (TSK gel Amido-80). The performances of three detectors (UV, CAD, and FLD) were compared. Under HILIC elution, CAD was five times more sensitive than UV. Acceptable linearity from 5.0 to 1000 pmol ($r^2 = 0.9982$) was also achieved. FLD for derivatized SGP was 10 times more sensitive than CAD. However, the derivatization method was more complex, and sample recovery was poor. HILIC-CAD was applied to acid-hydrolyzed SGP and two hydrolysis products, asialo-oligosaccharide and monosialo-oligosaccharide.

Guo and coworkers developed a general methodology to separate various carbohydrates using HILIC with either an ELSD or UV detector. The employed "click" column chemistry produced by bonding maltose to silica [149,150]. The carbohydrates studied ranged from galactooligosaccharides (neutral), carrageenan oligosaccharides (acidic), sodium alginate (acidic), and chitooligosaccharides (basic), up to higher molecular weight fructooligosaccharides. With water and acetonitrile mobile phase, galactooligosaccharides with different degrees of polymerization were separated. For the ionic carbohydrates, it was necessary to buffer the mobile phase pH with ammonium formate to improve peak shape. Their gradient method separated 40 peaks in 60 min, including fructooligosaccharides and high molecular weight oligosaccharides.

1.8   CONCLUSIONS AND FUTURE DIRECTIONS

Because of improvements in sensitivity, precision, and ease of use in the last ten years, aerosol-based detectors have arguably become the preferred alternative to UV for compounds lacking a chromophore. State-of-the-art detectors in this class are now capable of routinely achieving subnanogram detection limits, demonstrating precision for major component analysis within 2% RSD, and producing methods that are readily validated in a regulated environment. The major advancements that led to this were the commercial introductions of the CAD and CNLSD, which have also spurred improvements in ELSD technology.

Future advancements concern higher sensitivity for semivolatile and volatile compounds, broader linear and dynamic ranges, less susceptibility to gradient effects, and improved response uniformity. The ultimate goal is a robust, universal, mass sensitive detector capable of single-calibrant quantification.

REFERENCES

1. L. R. Snyder, J. J. Kirkland, and J. W. Dolan, *Introduction to Modern Liquid Chromatography*, 3rd edition, Wiley, Hoboken, 2010.
2. C. K. Lim and G. Lord, *Biol. Pharm. Bull.*, 25: 547 (2002).
3. R. Kostiainen and T. J. Kauppila, *J. Chromatogr. A*, 1216: 685 (2009).
4. J. A. Koropchak, L. E. Magnusson, M. Heybroek, S. Sadain, X. H. Yang, and M. P. Anisimov, in *Advances in Chromatography*, Brown, P. R., Grushka, E., Eds., Dekker, New York, 2000, pp. 275–314.
5. N. C. Megoulas and M. A. Koupparis, *Crit. Rev. Anal. Chem.*, 35: 301 (2005).
6. T. Vehovec and A. Obreza, *J. Chromatogr. A*, 1217: 1549 (2010).
7. P. Chaminade, in *Hyphenated and Alternative Methods of Detection in Chromatography*, Shalliker, R. A., Ed., CRC Press, Boca Raton, FL, 2012, pp. 145–160.
8. S. Almeling, D. Ilko, and U. Holzgrabe, *J. Pharm. Biomed. Anal.*, 69: 50 (2012).
9. R. Lucena, S. Cardenas, and M. Valcarcel, *Anal. Bioanal. Chem.*, 388: 1663 (2007).
10. W. O. Aruda, S. Walfish, and I. S. Krull, *LC GC N. Am.*, 26: 1032 (2008).
11. D. L. Ford and W. Kennard, *J. Oil Colour Chem. Assoc.*, 49: 299 (1966).
12. J. M. Charlesworth, *Anal. Chem.*, 50: 1414 (1978).
13. T. H. Mourey and L. E. Oppenheimer, *Anal. Chem.*, 56: 2427 (1984).
14. L. B. Allen and J. A. Koropchak, *Anal. Chem.*, 65: 841 (1993).

15. L. B. Allen, J. A. Koropchak, and B. Szostek, *Anal. Chem.*, 67: 659 (1995).
16. R. W. Dixon and D. S. Peterson, *Anal. Chem.*, 74: 2930 (2002).
17. B. Y. H. Liu and D. Y. H. Pui, *J. Aerosol Sci.*, 6: 249 (1975).
18. R. W. Dixon, *Bioanalysis*, 1: 1389 (2009).
19. P. H. Gamache, R. S. McCarthy, S. M. Freeto, D. J. Asa, M. J. Woodcock, K. Laws, and R. O. Cole, *LC GC Eur.*, 18: 345 (2005).
20. Y. Lv, S. C. Zhang, G. H. Liu, M. W. Huang, and X. R. Zhang, *Anal. Chem.*, 77: 1518 (2005).
21. G. M. Huang, Y. Lv, S. C. Zhang, C. D. Yang, and X. R. Zhang, *Anal. Chem.*, 77: 7356 (2005).
22. A. Stolyhwo, H. Colin, and G. Guiochon, *J. Chromatogr.*, 265: 1 (1983).
23. R. A. Mugele and H. D. Evans, *Ind. Eng. Chem.*, 43: 1317 (1951).
24. M. Dreux, M. Lafosse, and L. Morin-Allory, *LC GC Int.*, 9: 148 (1996).
25. M. Righezza and G. Guiochon, *J. Liq. Chromatogr.*, 11: 2709 (1988).
26. B. T. Mathews, P. D. Higginson, R. Lyons, J. C. Mitchell, N. W. Sach, M. J. Snowden, M. R. Taylor, and A. G. Wright, *Chromatographia*, 60: 625 (2004).
27. M. D. Barnes and V. K. La Mer, *J. Colloid Sci.*, 1: 79 (1946).
28. V. K. La Mer, *J. Phys. Colloid Chem.*, 52: 65 (1948).
29. D. Sinclair and V. K. La Mer, *Chem. Rev.*, 44: 245 (1949).
30. S. N. Timasheff, *J. Colloid Sci.*, 21: 489 (1966).
31. Y. Mengerink, H. C. J. De Man, and S. Van Der Wal, *J. Chromatogr.*, 552: 593 (1991).
32. Z. Cobb, P. N. Shaw, L. L. Lloyd, N. Wrench, and D. A. Barrett, *J. Microcolumn Sep.*, 13: 169 (2001).
33. L. E. Oppenheimer and T. H. Mourey, *J. Chromatogr.*, 323: 297 (1985).
34. P. Vandermeeren, J. Vanderdeelen, and L. Baert, *Anal. Chem.*, 64: 1056 (1992).
35. W. Guo, J. A. Koropchak, and C. Yan, *J. Chromatogr. A*, 849: 587 (1999).
36. S. K. Sadain and J. A. Koropchak, *J. Chromatogr. A*, 844: 111 (1999).
37. S. K. Sadain and J. A. Koropchak, *J. Liq. Chromatogr. Related Technol.*, 22: 799 (1999).
38. B. Szostek and J. A. Koropchak, *Anal. Chem.*, 68: 2744 (1996).
39. B. Szostek, J. Zajac, and J. A. Koropchak, *Anal. Chem.*, 69: 2955 (1997).
40. J. You and J. A. Koropchak, *J. Chromatogr. A*, 989: 231 (2003).
41. K. C. Lewis, D. M. Dohmeier, J. W. Jorgenson, S. L. Kaufman, F. Zarrin, and F. D. Dorman, *Anal. Chem.*, 66: 2285 (1994).
42. H. Bartz, H. Fissan, C. Helsper, Y. Kousaka, K. Okuyama, N. Fukushima, P. B. Keady, S. Kerrigan, S. A. Fruin, P. H. Mcmurry, D. Y. H. Pui, and M. R. Stolzenburg, *J. Aerosol Sci.*, 16: 443 (1985).
43. K. H. Ahn and B. Y. H. Liu, *J. Aerosol Sci.*, 21: 249 (1990).
44. J. A. Koropchak, C. L. Heenan, and L. B. Allen, *J. Chromatogr. A*, 736: 11 (1996).
45. R. C. Flagan, *Aerosol Sci. Tech.*, 28: 301 (1998).
46. M. Breysse, B. Claudel, L. Faure, M. Guenin, R. J. J. Williams, and T. Wolkenstein, *J. Catal.*, 45: 137 (1976).
47. Y. F. Zhu, J. J. Shi, Z. Y. Zhang, C. Zhang, and X. R. Zhang, *Anal. Chem.*, 74: 120 (2002).
48. X. O. Cao, Z. Y. Zhang, and X. R. Zhang, *Sensors Actuators B Chem.*, 99: 30 (2004).
49. R. S. Gao, A. E. Perring, T. D. Thornberry, A. W. Rollins, J. P. Schwarz, S. J. Ciciora, and D. W. Fahey, *Aerosol Sci. Technol.*, 47: 137 (2013).
50. J. P. Hutchinson, J. F. Li, W. Farrell, E. Groeber, R. Szucs, G. Dicinoski, and P. R. Haddad, *J. Chromatogr. A*, 1218: 1646 (2011).
51. L. M. Nair and J. O. Werling, *J. Pharm. Biomed. Anal.*, 49: 95 (2009).
52. R. G. Ramos, D. Libong, M. Rakotomanga, K. Gaudin, P. M. Loiseau, and P. Chaminade, *J. Chromatogr. A*, 1209: 88 (2008).
53. H. Y. Eom, S. Y. Park, M. K. Kim, J. H. Suh, H. Yeom, J. W. Min, U. Kim, J. Lee, J. R. Youm, and S. B. Han, *J. Chromatogr. A*, 1217: 4347 (2010).

54. J. Shaodong, W. J. Lee, J. W. Ee, J. H. Park, S. W. Kwon, and J. Lee, *J. Pharm. Biomed. Anal.*, 51: 973 (2010).
55. L. Wang, W. S. He, H. X. Yan, Y. Jiang, K. S. Bi, and P. F. Tu, *Chromatographia*, 70: 603 (2009).
56. R. D. Cohen, Y. Liu, and X. Y. Gong, *J. Chromatogr. A*, 1229: 172 (2012).
57. N. Vervoort, D. Daemen, and G. Torok, *J. Chromatogr. A*, 1189: 92 (2008).
58. R. Vogel, R. D. Cohen, and P. Diamandopoulos, *Trends Chromatogr.*, 4: 25 (2008).
59. J. M. Cintron and D. S. Risley, *J. Pharm. Biomed. Anal.*, 78–79: 14 (2013).
60. A. Morales, S. Marmesat, M. C. Dobarganes, G. Marquez-Ruiz, and J. Velasco, *J. Chromatogr. A*, 1254: 62 (2012).
61. W. J. Deng, X. N. Zhang, Z. P. Sun, J. L. Yin, Z. Zhou, L. P. Han, and S. J. Zhao, *Chromatographia*, 75: 629 (2012).
62. Y. H. Wei, L. W. Qi, P. Li, H. W. Luo, L. Yi, and L. H. Sheng, *J. Pharm. Biomed. Anal.*, 45: 775 (2007).
63. R. Rombaut, K. Dewettinck, and J. Van Camp, *J. Food Compos. Anal.*, 20: 308 (2007).
64. S. Fekete, K. Ganzler, and J. Fekete, *J. Chromatogr. A*, 1217: 6258 (2010).
65. J. Olsovska, Z. Kamenik, and T. Cajthaml, *J. Chromatogr. A*, 1216: 5774 (2009).
66. S. Inagaki, J. Z. Min, and T. Toyo'oka, *Biomed. Chromatogr.*, 21: 338 (2007).
67. C. Schönherr, S. Touchene, G. Wilser, R. Peschka-Suss, and G. Francese, *J. Chromatogr. A*, 1216: 781 (2009).
68. M. Lísa, F. Lynen, M. Holcapek, and P. Sandra, *J. Chromatogr. A*, 1176: 135 (2007).
69. J. P. Hutchinson, T. Remenyi, P. Nesterenko, W. Farrell, E. Groeber, R. Szucs, G. Dicinoski, and P. R. Haddad, *Anal. Chim. Acta*, 750: 199 (2012).
70. K. Zhang, L. L. Dai, and N. P. Chetwyn, *J. Chromatogr. A*, 1217: 5776 (2010).
71. A. Espada and A. Rivera-Sagredo, *J. Chromatogr. A*, 987: 211 (2003).
72. K. Petritis, H. Dessans, C. Elfakir, and M. Dreux, *LC GC Eur.*, 15: 98 (2002).
73. Z. Huang, M. A. Richards, Y. Zha, R. Francis, R. Lozano, and J. Ruan, *J. Pharm. Biomed. Anal.*, 50: 809 (2009).
74. T. Teutenberg, J. Tuerk, M. Holzhauser, and T. K. Kiffmeyer, *J. Chromatogr. A*, 1119: 197 (2006).
75. M. Righezza and G. Guiochon, *J. Liq. Chromatogr.*, 11: 1967 (1988).
76. C. Brunelli, T. Gorecki, Y. Zhao, and P. Sandra, *Anal. Chem.*, 79: 2472 (2007).
77. S. Lane, B. Boughtflower, I. Mutton, C. Paterson, D. Farrant, N. Taylor, Z. Blaxill, C. Carmody, and P. Borman, *Anal. Chem.*, 77: 4354 (2005).
78. V. L. Cebolla, L. Membrado, J. Vela, and A. C. Ferrando, *J. Chromatogr. Sci.*, 35: 141 (1997).
79. J. P. Hutchinson, J. F. Li, W. Farrell, E. Groeber, R. Szucs, G. Dicinoski, and P. R. Haddad, *J. Chromatogr. A*, 1217: 7418 (2010).
80. S. H. Hilal, S. W. Karickhoff, and L. A. Carreira, *Verification and Validation of the SPARC Model,* U.S. Environmental Protection Agency, Washington, DC, 2003.
81. A. W. Squibb, M. R. Taylor, B. L. Parnas, G. Williams, R. Girdler, P. Waghorn, A. G. Wright, and F. S. Pullen, *J. Chromatogr. A*, 1189: 101 (2008).
82. C. Crafts, B. Bailey, P. Gamache, X. Liu, and I. Acworth, in *Applications of Ion Chromatography for Pharmaceutical and Biological Products*, Bhattacharyya, L., Rohrer, J. S., Eds., Wiley, Hoboken, NJ, 2012, pp. 221–236.
83. P. Wipf, S. Werner, I. A. Twining, and C. Kendall, *Chirality*, 19: 5 (2007).
84. D. W. Kou, G. Manius, S. D. Zhan, and H. P. Chokshi, *J. Chromatogr. A*, 1216: 5424 (2009).
85. G. Guiochon, A. Moysan, and C. Holley, *J. Liq. Chromatogr.*, 11: 2547 (1988).
86. S. Nukiyama and Y. Tanasawa, *Trans. Soc. Mech. Eng., Tokyo*, 4: 86 (1938).
87. T. Gorecki, F. Lynen, R. Szucs, and P. Sandra, *Anal. Chem.*, 78: 3186 (2006).
88. A. de Villiers, T. Gorecki, F. Lynen, R. Szucs, and P. Sandra, *J. Chromatogr. A*, 1161: 183 (2007).
89. K. Mojsiewicz-Pienkowska, *Crit. Rev. Anal. Chem.*, 39: 89 (2009).

90. L. L. Fang, M. Wan, M. Pennacchio, and J. M. Pan, *J. Comb. Chem.*, 2: 254 (2000).
91. C. E. Kibbey, *Mol. Divers.*, 1: 247 (1995).
92. B. H. Hsu, E. Orton, S. Y. Tang, and R. A. Carlton, *J. Chromatogr. B*, 725: 103 (1999).
93. H. E. Fries, C. A. Evans, and K. W. Ward, *J. Chromatogr. B*, 819: 339 (2005).
94. L. Barantin, A. Le Pape, and S. Akoka, *Magn. Reson. Med.*, 38: 179.
95. A. Stojanovic, M. Lämmerhofer, D. Kogelnig, S. Schiesel, M. Sturm, M. Galanski, R. Krachler, B. K. Keppler, and W. Lindner, *J. Chromatogr. A*, 1209: 179 (2008).
96. I. Sinclair and I. Charles, *J. Biomol. Screen.*, 14: 531 (2009).
97. K. Petritis, C. Elfakir, and M. Dreux, *J. Chromatogr. A*, 961: 9 (2002).
98. A. A. Adoubel, S. Guenu, C. Elfakir, and M. Dreux, *J. Liq. Chromatogr. Related Technol.*, 23: 2433 (2000).
99. K. Takahashi, S. Kinugasa, M. Senda, K. Kimizuka, K. Fukushima, T. Matsumoto, Y. Shibata, and J. Christensen, *J. Chromatogr. A*, 1193: 151 (2008).
100. K. Takahashi, S. Kinugasa, R. Yoshihara, A. Nakanishi, R. K. Mosing, and R. Takahashi, *J. Chromatogr. A*, 1216: 9008 (2009).
101. A. Joseph and A. Rustum, *J. Pharm. Biomed. Anal.*, 51: 521 (2010).
102. K. Stypulkowska, A. Blazewicz, Z. Fijalek, and K. Sarna, *Chromatographia*, 72: 1225 (2010).
103. U. Holzgrabe, C. J. Nap, N. Kunz, and S. Almeling, *J. Pharm. Biomed. Anal.*, 56: 271 (2011).
104. B. Forsatz and N. H. Snow, *LC GC N. Am.*, 25: 960 (2007).
105. P. R. Haddad, P. N. Nesterenko, and W. Buchberger, *J. Chromatogr. A*, 1184: 456 (2008).
106. S. P. Chen, T. Huang, and S. G. Sun, *J. Chromatogr. A*, 1089: 142 (2005).
107. P. R. Haddad, P. E. Jackson, and M. J. Shaw, *J. Chromatogr. A*, 1000: 725 (2003).
108. J. D. Lamb, D. Simpson, B. D. Jensen, J. S. Gardner, and Q. P. Peterson, *J. Chromatogr. A*, 1118: 100 (2006).
109. R. Slingsby and R. Kiser, *TrAC Trends Anal. Chem.*, 20: 288 (2001).
110. P. R. J. Haddad and P. E. Jackson, *Ion Chromatography: Principles and Applications*, Elsevier Science, Amsterdam, 1990.
111. M. D. Lantz, D. S. Risley, and J. A. Peterson, *J. Liq. Chromatogr. Related Technol.*, 20: 1409 (1997).
112. C. McClintic, D. M. Remick, J. A. Peterson, and D. S. Risley, *J. Liq. Chromatogr. Related Technol.*, 26: 3093 (2003).
113. B. W. Pack and D. S. Risley, *J. Chromatogr. A*, 1073: 269 (2005).
114. D. S. Risley and B. W. Pack, *LC GC N. Am.*: 82 (2006).
115. P. Hemstrom and K. Irgum, *J. Sep. Sci.*, 29: 1784 (2006).
116. B. Dejaegher, D. Mangelings, and Y. V. Heyden, *J. Sep. Sci.*, 31: 1438 (2008).
117. C. Crafts, B. Bailey, M. Plante, and I. Acworth, *J. Chromatogr. Sci.*, 47: 534 (2009).
118. C. Crafts, B. Bailey, and I. Acworth, *LC GC N. Am.*: 43 (2010).
119. X. D. Liu and C. A. Pohl, *J. Sep. Sci.*, 33: 779 (2010).
120. W. W. Christie, *J. Lipid Res.*, 26: 507 (1985).
121. R. A. Moreau, *Lipid Technol.*, 21: 191 (2009).
122. G. Kielbowicz, D. Smuga, W. Gladkowski, A. Chojnacka, and C. Wawrzenczyk, *Talanta*, 94: 22 (2012).
123. A. Hazotte, D. Libong, A. Matoga, and P. Chaminade, *J. Chromatogr. A*, 1170: 52 (2007).
124. C. Merle, C. Laugel, P. Chaminade, and A. Baillet-Guffroy, *J. Liq. Chromatogr. Related Technol.*, 33: 629 (2010).
125. R. A. Moreau, in *HPLC of Acyl Lipids*, Lin, J.-T., McKeon, T. A., Eds., H.N.B., New York, 2005, pp. 93–116.
126. F. S. Deschamps, P. Chaminade, D. Ferrier, and A. Baillet, *J. Chromatogr. A*, 928: 127 (2001).
127. M. Holcapek, M. Lisa, P. Jandera, and N. Kabatova, *J. Sep. Sci.*, 28: 1315 (2005).
128. R. Diaz-Lopez, D. Libong, N. Tsapis, E. Fattal, and P. Chaminade, *J. Pharm. Biomed. Anal.*, 48: 702 (2008).

129. R. G. Strickley, *Pharm. Res.*, 21: 201 (2004).
130. T. Alfredson and M. Zeller, *Drug Discov. Today Technol.*, 9: e71 (2012).
131. M. Kuentz, *Drug Discov. Today Technol.*, 9: e97 (2012).
132. Y. S. R. Krishnaiah, *J. Bioequiv. Availab.*, 2: 028 (2010).
133. A. Nozawa, T. Ohnuma, and S. Tatsuya, *Analyst*, 101: 543 (1976).
134. N. H. Anderson and J. Girling, *Analyst*, 107: 836 (1982).
135. S. Fekete, K. Ganzler, and J. Fekete, *J. Pharm. Biomed. Anal.*, 52: 672 (2010).
136. J. M. Harris and R. B. Chess, *Nat. Rev. Drug Discov.*, 2: 214 (2003).
137. P. N. Nesterenko and V. I. Savel'ev, *J. Anal. Chem. USSR*, 45: 819 (1990).
138. M. Korosec, J. Bertoncelj, A. P. Gonzales, U. Kropf, U. Golob, and T. Golob, *Acta Aliment. Hung.*, 38: 459 (2009).
139. C. Borromei, A. Cavazza, C. Merusi, and C. Corradini, *J. Sep. Sci.*, 32: 3635 (2009).
140. K. Guilloux, I. Gaillard, J. Courtois, B. Courtois, and E. Petit, *J. Agr. Food Chem.*, 57: 11308 (2009).
141. C. Antonio, T. Larson, A. Gilday, I. Graham, E. Bergstroem, and J. Thomas-Oates, *J. Chromatogr. A*, 1172: 170 (2007).
142. J. C. Linden and C. L. Lawhead, *J. Chromatogr. A*, 105: 125 (1975).
143. D. Asa, *Am. Lab.*, 38: 16 (2006).
144. C. R. Mitchell, Y. Bao, N. J. Benz, and S. H. Zhang, *J. Chromatogr. B*, 877: 4133 (2009).
145. B. Godin, R. Agneessens, P. A. Gerin, and J. Delcarte, *Talanta*, 85: 2014 (2011).
146. S. Hase, T. Ibuki, and T. Ikenaka, *J. Biochem. (Tokyo)*, 95: 197 (1984).
147. A. Kon, K. Takagaki, H. Kawasaki, T. Nakamura, and M. Endo, *J. Biochem. (Tokyo)*, 110: 132 (1991).
148. K. Takagaki, T. Nakamura, H. Kawasaki, A. Kon, S. Ohishi, and M. Endo, *J. Biochem. Biophys. Methods*, 21: 209 (1990).
149. Z. M. Guo, A. W. Lei, X. M. Liang, and Q. Xu, *Chem. Commun.*: 4512 (2006).
150. Z. M. Guo, A. W. Lei, Y. P. Zhang, Q. Xu, X. Y. Xue, F. F. Zhang, and X. M. Liang, *Chem. Commun.*: 2491 (2007).

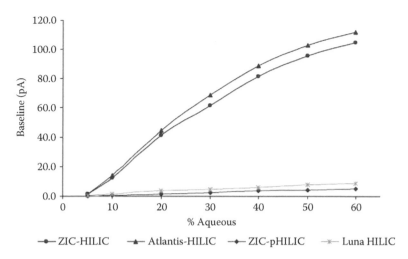

COLOR FIGURE 1.7 Column bleed from four different HILIC columns as a function of aqueous component of mobile phase. Detector = Corona® CAD; column temperature = ambient; mobile phase buffer = 0.1 M ammonium acetate, pH 7.0; organic solvent = acetonitrile; flow rate = 1.0 mL/min. (Reprinted from Z. Huang, M. A. Richards, Y. Zha, R. Francis, R. Lozano, and J. Ruan, *J. Pharm. Biomed. Anal.*, 50: 809, 2009, with permission from Elsevier. Copyright 2009.)

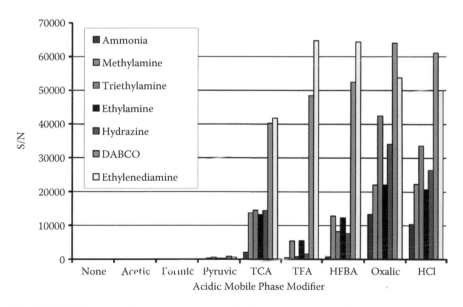

COLOR FIGURE 1.18 Improvement in volatile base response by addition of volatile acidic mobile phase modifier. TCA, trichloroacetic acid; TFA, trifluoroacetic acid; HFBA, heptafluorobutyric acid; DABCO, 1,4-diazabicyclo[2.2.2]octane.

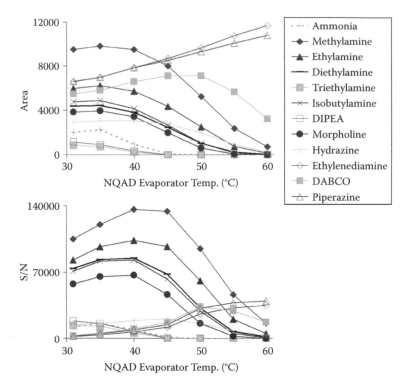

COLOR FIGURE 1.20 Effect of CNLSD response versus evaporator temperature for volatile bases, each at a concentration of 0.1 mg/mL. Mobile phase = 0.2 mM HCl in trifluoroacetic acid/acetonitrile/water (0.04:60:40, v/v/v) used for monoprotic bases and mobile phase = trifluoroacetic acid/acetonitrile/water (0.4:70:30, v/v/v) used for polyprotic bases; column = zic-*p*HILIC; flow rate = 1.0 mL/min; injection volume = 5.0 μL; column temperature = 40°C. (Reprinted from R. D. Cohen, Y. Liu, and X. Y. Gong, *J. Chromatogr. A*, 1229: 172, 2012, with permission from Elsevier. Copyright 2012.)

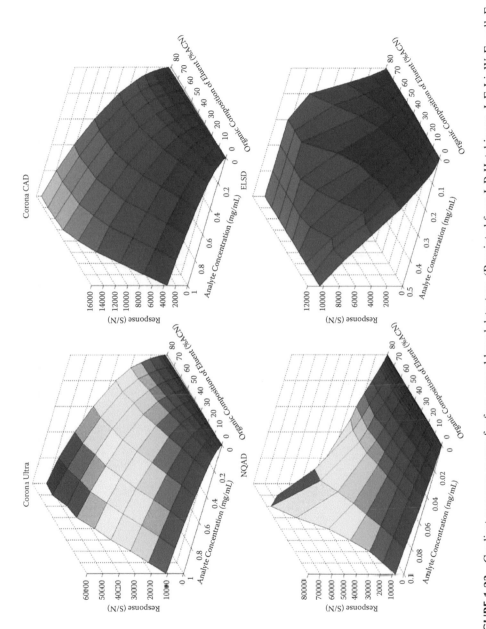

COLOR FIGURE 1.23 Gradient response curves for four aerosol-based detectors. (Reprinted from J. P. Hutchinson, J. F. Li, W. Farrell, E. Groeber, R. Szucs, G. Dicinoski, and P. R. Haddad, *J. Chromatogr. A*, 1218: 1646, 2011, with permission from Elsevier. Copyright 2011.)

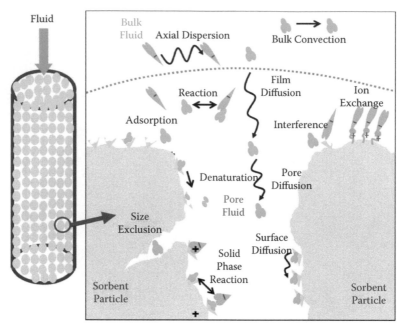

COLOR FIGURE 5.2 Various reactions and mass transfer mechanisms considered in VERSE.

2 Integration of Analytical Techniques for Protein Biomarker Assays

Jennifer L. Colangelo

2.1 INTRODUCTION

Effective biomarkers are among the most sought after tools in the area of human health. They are used to diagnose diseases, monitor drug efficacy and safety, aid in drug selection, and monitor overall general health. Many different endpoints are considered biomarkers: Elevated blood pressure can indicate the risk of stroke, elevated liver enzymes can indicate liver injury, and genetic signatures can indicate predisposition to certain diseases. Protein levels and activity are often used in health assessments as biomarkers [1]. For example, the activity level of alanine transaminase (ALT) in the blood indicates how well the liver is functioning, and elevated levels of C-reactive protein (CRP) can indicate inflammatory disease or response. Recently, many novel proteins are being proposed as potential markers of disease states [2,3]. Therefore, the demand for robust, quantitative assays has increased as more and more proteins are being investigated and deployed as biomarkers. This chapter focuses on protein and peptide biomarkers that are chromatographic endpoints measured in biological fluids and tissues, such as serum, urine, and tissue biopsies. Table 2.1 highlights a number of existing quantitative protein/peptide assays that employ chromatography.

TABLE 2.1

Quantitative Protein Assays that Employ Chromatography

Analyte	Platform	Chromatography	Matrix	Application	Reference
Angiotensin II	LC-MS	Reversed phase	Plasma	Vascular health, angiotensin-converting enzyme (ACE) administration	22
C-reactive protein (CRP)	LC-MS	Reversed phase	Plasma	Inflammation, atherosclerosis, cardiovascular	23, 24
Ceruloplasmin (CP)	LC-MS	Reversed phase	Dried blood spot	Wilson's disease	25
Interleukin 33 (IL-33)	LC-MS	Reversed phase	Plasma	Immune response (allergy and autoimmune diseases)	26
Myocin light chain 1 (Myl 1/3)	LC-MS	Reversed phase	Serum	Cardiovascular	27, 28
Myoglobin	LC-MS	Reversed phase	Serum	Cardiovascular	29
N-Terminal pro-brain natriuretic peptide (NTproBNP)	LC-MS	Reversed phase	Serum	Cardiovascular (cardiac hypertrophy)	28, 30
Procollagen type I N-terminal propeptide (P1NP)	LC-MS	Reversed phase	Serum	Bone health	31
Prostate-specific antigen (PSA)	LC-MS	Strong cation exchange (SCX) and reversed phase	Serum	Prostate health	32
Troponin I	LC-MS	Reversed phase	Plasma/serum	Cardiovascular	26, 29

Establishing new protein biomarkers could have a broad, positive impact on the diagnosis and detection of various disease areas, especially those areas for which none exist. For example, no biomarkers currently exist to diagnose or predict glomerular injury of the kidney [4,5]. Glomerular injury and disease are known manifestations during a number of fairly common metabolic and cardiovascular diseases, such as diabetes and hypertension, as well as inflammatory diseases, such as lupus [6]. Biomarkers of overall kidney function, such as BUN (serum [blood] urea nitrogen) and creatinine, are often used yet typically do not demonstrate any measurable changes until the kidney is significantly damaged [2,7]. Being able to detect this type of injury earlier in the process would be beneficial to these patients. Another example is the need to better detect adverse drug reactions in the liver during the drug development process. Liver injury is the most frequent cause of safety-related drug withdrawals [3]. Serum levels of ALT activity are the gold standard in the clinic to indicate hepatotoxic effects, and other liver-specific enzymes are also used to gain insight to liver function [8]. The difficulty in drug development occurs when no signs of liver injury manifest in preclinical studies but are observed in the clinical setting or when the drug hits a larger population. Better indicators of the potential for liver toxicity in humans are needed in the preclinical assessments for the development of safer drugs [3,8,9].

Development of a new biomarker for use in the clinical setting can be difficult and time consuming [1,10]. First, a disease-specific biomarker must be identified. Complex and challenging experiments, such as discovery-based proteomics, are typically employed to identify potential proteins that change with disease or injury [2,11–15]. While very powerful, the widespread use of proteomic tools in recent years has created a vast number of potential protein biomarkers for various disease states, yet following up on the potential of each protein is a daunting task. Discovery proteomic experiments typically produce lists containing hundreds of proteins as potential biomarkers, so other tools, such as pathway analyses and targeted proteomics, are utilized to narrow the number of potential biomarkers [16–18].

Once potential biomarkers are identified, a suitable assay to measure them is required. The proteomic assays used for discovery do not make the best tools for further investigation into biomarker performance. Often, proteomic techniques measure the relative concentrations of hundreds of proteins to each other. Because the assays are not specific to a single protein or group of proteins, the data can be skewed or difficult to reproduce. In addition, these proteomic analyses often require a long run time, which is appropriate for profiling a small number of samples but not for routine analyses of hundreds or thousands of samples. A more specific assay is preferable to characterize the protein biomarker and its biological relevance [1]. Thus, assays must be developed to obtain data for the proteins of interest.

Biomarker assays for proteins are often built from antibody technologies, which have benefits and disadvantages. Assays, such as enzyme-linked immunosorbent assays (ELISAs), capitalize on the binding specificity of the antibody to the protein for detection of the protein. Most assays are straightforward and easy to execute. In the clinical setting, many antibody-based assays are established on routine clinical chemistry analyzers, for which kits and reagents can be purchased to perform the analyses. In a research environment, assays are often performed by ELISA or radioimmunoassay (RIA) and can be multiplexed on readers designed for that pur-pose. The challenge comes in finding reagents for novel biomarkers or for multiple preclinical species [19]. Sometimes, reagents may not be robust, may demonstrate lot-to-lot differences, or may not be available on a consistent basis. For these reasons, other technologies, such as chromatography and mass spectrometry (MS), have been employed to fill this gap [20]. For liquid chromatographic (LC)/MS assays, specific antibodies are not necessary to develop a quantitative assay; sensitivity is on par with antibody-based platforms; these platforms are now being found in clinical hospital laboratories and other similar settings. These technologies provide another advan-tage in that they are often used for proteomic experiments. So, for those protein biomarkers identified through proteomics experiments, starting conditions for assay development can be identified, streamlining the process [21].

The evaluation process for a new protein biomarker includes the technical valida-tion of the assay to provide confidence in the assay and the biological evaluation of the biomarker. Standard evaluations of assay performance can include many different assessments that are relevant to the technique or sample, such as inter- and intraas-say performance, carryover, interferences, and linearity. To qualify the biomarker in regard to its biological performance, several investigations must be performed: Reference ranges must be established; performance in different disease states and populations must be evaluated; and the effect of variables, such as sex and age, must be considered [1,3]. Understanding how the biomarker behaves in all types of settings is important for proper use and interpretation of results. Then, the biomarker must be rigorously tested across laboratories and in a large population, in both the normal and disease states [1,3].

The focus of this chapter is on the integration of analytical techniques for protein biomarker assays, specifically the integration of LC and MS with relevant analytical tools. Discussions include chromatography applications, how scale should be con-sidered, and integration of sample-handling and -labeling techniques. Some of these discussions could be applied to biomarker assays for smaller molecules or other bio-logical molecules due to the nature of assay development for biomarkers. Here, the focus is on protein applications.

2.2   THE ROLE OF CHROMATOGRAPHY IN PROTEIN BIOMARKER ASSAYS

As tools for biomarker discovery are increasingly being used to search for new pro-tein biomarkers in the proteome, researchers need to be able to detect and quantitate proteins and peptides at low nanomolar and picomolar levels. In some cases, scien-tists are taking new looks at old biomarkers, some of which are in the micromolar range [33]. For most cases, biological fluids and tissues contain such a wide variety of constituents (proteins, antibodies, and endogenous metabolites) that chromato-graphic separation is a critical step in protein assay development due to its ability to separate complex mixtures.

It can be difficult to separate the protein of interest from all these other molecu-lar species. To facilitate this separation, often the protein mixture is enzymatically digested at some point in the process, creating a complex mixture of peptides, since fractionating peptides is often easier than proteins. The challenge of analyzing a digested peptide mixture is that the complexity of the mixture has been exponentially increased, and separation of a surrogate peptide without interference from the other components must occur. To minimize interference, these separations must be conducted in an efficient manner, often relying on chromatography to achieve the necessary resolution between various peptides.

2.2.1 MODES OF SEPARATIONS

The choice of chromatography that will be used in the assay is also critical to assay success. The most common choices for protein and peptide applications are reversed-phase, affinity, ion exchange, and size exclusion chromatography. Due to the complex nature of samples, often more than one mode is used to aid in reducing the complexity of the mixture, and some of these may be used off line as part of sample cleanup. Separating the peptide sufficiently from the other components for adequate detection is one of the key requirements of the assay.

The most commonly selected mode of chromatography for the quantitation of peptides is reversed phase. Samples that have been enzymatically digested to a mixture of small peptides can be successfully resolved with reversed-phase chromatography. Standard column technologies with C18 phases are commonly used, and these analyses are considered relatively routine. Reversed-phase chromatography is also very amenable to pairing with various forms of detection, especially MS, as illustrated in the number of biomarker assays utilizing reversed-phase LC-MS in Table 2.1. Typically, standard linear gradient separation is deployed in reversed-phase mode since it provides a good basis for eluting many peptides due to its higher peak capacity than isocratic separations.

Although reversed phase is the most popular mode of chromatography used for quantitative protein biomarker assays, other types of chromatography can be used and provide options that may be advantageous based on the properties of the protein or peptide. Ion exchange chromatography is beneficial if separation based on the net surface charge provides greater resolution and is a technique often employed for the separation of phosphopeptides [34]. The mobile phases used for ion exchange contain salts that are not compatible with mass spectrometers, interfering with ionization. Therefore, this technique is often used as a precursor to another, more compatible, technique, such as reversed-phase chromatography, as exemplified in the assay developed for prostate-specific antigen (PSA) (Table 2.1). Affinity techniques may be chosen based on unique properties of the analyte of interest or based on labeling techniques [35]. Glycoproteins and glycopeptides can be fractionated from complex samples with lectin affinity chromatography since lectins have the ability to bind to specific carbohydrate residues. Immobilized metal affinity chromatography is another technique often applied to the analysis of phosphopeptides [36]. Often, affinity techniques based on an antibody that binds to the analyte of interest are utilized off line in the sample preparation, which is discussed in a further section. Finally, size exclusion chromatography is often used for the fractionation of proteins in complex mixtures prior to other proteomic analyses [37,38]. With regard to quantitative protein and peptide assays, this separation technique may be used off line as well to aid in the fractionation of the sample [39].

2.2.2   LC System Selection

In determining the assay platform, the type of chromatography system influences the selectivity and sensitivity. Conventional high-performance liquid chromatographic (HPLC) systems are commonly found in laboratories and can be interfaced to a wide variety of detectors. Most vendors offer robust systems with the same features, such as column heaters, autosamplers, and an array of detectors. Various techniques can be used on conventional HPLC systems to increase assay efficiency: elevated temperatures, faster flow rates, and smaller columns. Flow rate of the system is an important consideration depending on the detector that is to be interfaced to the system in that the two need to be compatible.

Recently, ultraperformance liquid chromatographic (UPLC) commercial systems have emerged on the market, offering several advantages over conventional HPLC systems [40]. The first commercial UPLC system was available in 2004, so even though these systems have not been readily available long, they have become rapidly integrated into protein quantitative assays. UPLC systems operate at a greater pressure range, providing a system that can handle columns with sub-2-μm particle columns. Since these columns have a smaller diameter, researchers can use low-volume injections and reduce the amount of solvent used. Due to the reduction in particle dispersion on the column, separations are able to achieve higher resolution, increased peak capacity, and faster run times, as illustrated in Figure 2.1. Increased peak capacity is critical when trying to separate complex mixtures of peptides.

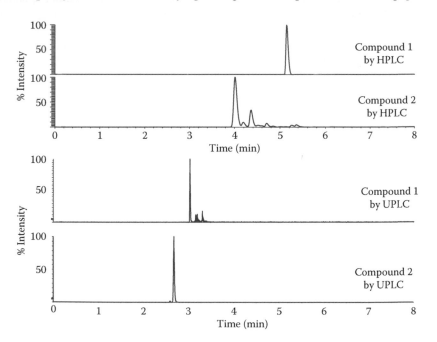

FIGURE 2.1   Comparison between HPLC and UPLC. The traces illustrate the increase in peak resolution and decrease in run time for two compounds when analyzed on a UPLC system in comparison to a HPLC system.

Overall, UPLC provides two to five times more sensitivity than HPLC. Peak widths are drastically reduced in comparison to HPLC, so one difficulty is making sure the detector is operating at a fast enough sampling rate as the peak elutes to capture a sufficient number of data points. Mass spectrometers are prime candidates to interface with UPLC systems due their fast scanning rates.

Nanoflow LC systems provide even greater sensitivity and enable extremely low sample volumes. They also consume an extremely small amount of solvent that reduces the cost for an individual run in comparison to other systems. The low flow rates used in these systems can complicate delay volume issues, make troubleshooting of leakage and flow blockage difficult, and result in long run times. In addition, reducing dead volumes in both switching valves and downstream from the LC column is critical since it can contribute to band broadening. Finally, the nanoelectrospray emitters used for interfacing to a mass spectrometer are often made of glass capillaries with orifices of only 1 to 10 mm in diameter. The difficulty with such small orifices is that complex biological samples can often clog and contaminate the tip, thus losing the remaining run and increasing instrument downtime. Nanoflow systems are sometimes used for quantitative protein assays yet are more commonly used as a discovery proteomics tool [24,29].

2.2.3 Additional Considerations

To develop a successful LC assay, other factors must be considered. For example, the quality and stability of the packed stationary phase in the column are critical. Finding columns that are robust over their expected lifespan and have limited lot-to-lot variability is necessary to make the assay viable in both inter- and intralaboratory assessments. A wide variety of column chemistries is available from multiple manufacturers; these chemistries are compatible with HPLC, UPLC, and nano LC systems [41]. Of course, the quality of mobile phase reagents is important, with the use of HPLC-grade solvents essential. The mobile phases and their stock solutions can introduce contaminates into the LC system, reducing the quality of the assay and results. Finally, routine service of LC systems is essential to maintain optimal assay performance since components, like seals, pumps, and injector ports, need preventive maintenance to ensure the system is operating properly.

Since biological samples, such as plasma, contain an abundance of proteins and peptides, multidimensional LC should be considered for analysis, especially for the enrichment of lower-abundance proteins and peptides. These techniques are mostly used in proteomic experiments when the maximum separation of peptides in the mixture needs to be achieved, yet these runs can be time consuming and not amenable to efficient, robust assays for routine use. Many different configurations for both the instrumentation setup and for the modes of chromatography can be used in these experiments [42]. Separations can be conducted off line, so that fractions are collected from the eluent of the one dimension and then injected into the next dimension for further separation. The article describing the quantitative assay for PSA demonstrated that performing strong cation exchange chromatography off line, collecting fractions of interest, and then performing the LC-MS analysis reduced the number of peptides in the sample and provided the potential for a lower limit of detection [32].

Separations can also be conducted online with seamless multidimensional systems; the eluent from the one dimension directly enters the next dimension. The separation can be performed so that only a relevant portion of the effluent enters the next dimension, or so that the entire sample goes through all dimensions. Transfer between the dimensions should be considered since the separation in the sample can be diminished due to mixing as the sample enters the next dimension.

For multidimensional chromatography to work, the compatibility of solvents between the separations also must be considered. Most two-dimensional separations for protein and peptide analyses employ either reversed-phase chromatography in both dimensions or use size exclusion or ion exchange chromatography in the first dimension with reversed-phase chromatography in the second [27,28,38]. The solvent system for the final chromatographic technique must be compatible with the detection method and not result in interferences. For example, the salts utilized in the mobile phases for some chromatographic applications produce adducts or interfere with ionization in the mass spectrometer, which is why reversed-phase chromatography is typically the mode of choice for this interface. The most basic two-dimensional separation employing the reversed phase is the use of a C18 trapping column in line with a C18 column, as is used in the LC-MS assays for myoglobin light chain and NTproBNP (N-terminal pro-brain natriuretic peptide) [27,28]. The sample is loaded onto the trapping column, allowing the contaminants, like salts and other impurities, to elute to waste before introducing the sample to the analytical column.

2.3   INTEGRATING ANALYTICAL PLATFORMS

Because of the complex nature of biological samples, the choice of the analytical platform or detector to interface with the chromatography system is critical to achieving the goals of the assay. A combination of selectivity and sensitivity tends to be the largest driver for the selection of the interface. Therefore, the selectivity and sensitivity needs of the assay need to be evaluated and matched to the appropriate platform. As discussed, many protein and peptide biomarker candidates are at low levels relative to the rest of the proteome, so sensitivity is a necessity. Selectivity is important in that many proteins and peptides have similar characteristics, may be members of an entire family of related proteins, or may have many isomers, for which biological activity may differ. These proteins and peptides can be distinguished from each other by measuring mass differences, monitoring unique fragment ions, or using special tags for detection. Of course, the chromatographic system enables separation, yet peptides with small differences in amino acid sequence, such as the difference between an alanine and a glycine, are difficult to separate. Detection schemes that can distinguish these small differences are often necessary.

2.3.1   Mass Spectrometry

The most common interface for quantitative protein biomarker assays is the mass spectrometer. When interfaced to a chromatographic system, this platform provides both selectivity and sensitivity for the quantitation of proteins and peptides [43]. Typically, selective reaction monitoring (SRM) scanning on triple-quadrupole mass spectrometers is employed. For SRM analysis, electrospray ionization (ESI) is used so that the LC eluent is ionized with an electrospray needle where the peptides are desolvated and ionized into the gas phase before entering the mass spectrometer. In a triple quadrupole, the first quadrupole of the mass spectrometer is used to filter the peptide based on the intact mass. The second quadrupole is where the peptide is fragmented using collision-induced dissociation, and the last quadrupole is used to filter the unique fragment ion of the peptide. This process is similar in other instrument configurations and enables researchers to confidently identify and quantitate the analyte of interest.

Proteins can be quantitated either from a single peptide or from a set of peptides, using multiple transitions for each peptide. Often, three peptides at minimum are monitored in the experiment, and three transitions for each peptide are observed. The accurate mass measurements enable small mass differences to be detected between peptides that co-elute, and unique signature fragment ions boost confidence that the correct protein is being identified. The quantitation can be conducted by an average of the signals or based on a single signal, using the others as confirmatory results. Experimental results can be used to determine which method is best for the protein. For example, some peptides or transitions may not ionize as reproducibly as others, making them less than ideal for performing quantitation.

Since the length and sequence of individual proteins are variable, identification of three peptides, each with three transitions, is not always achievable. The number of peptides can be limited due the number of peptides that are unique to the proteome and the size of the peptides produced from the enzymatic digestion (either too small or too large). Peptides that are too long pose a few challenges. The multiply charged ions formed in the mass spectrometer that are in the available mass range do not ionize efficiently. Also, in making standard peptides, the longer the peptide, the more difficult it is to synthesize. For example, the quantitative assay for procollagen type I N-terminal propeptide (P1NP) uses multiple transitions of a single peptide for quantitation, shown in Figure 2.2. The intact P1NP protein is made up of two procollagen type 1 alpha 1 chains and one procollagen type 1 alpha 2 chain for a total of 321 amino acids with a molecular mass of 33 kDa. On trypsin digestion, only one peptide had ideal qualities: good size, unique to the protein, and no potential modification sites that would complicate analysis [31].

Different types of configurations exist for the mass spectrometers, with some more preferable for protein biomarker assays. The three platforms most commonly used are the triple quadrupole, the ion trap, and the quadrupole ion trap, with the majority of assays, such as those listed in Table 2.1, utilizing triple-quadrupole mass spectrometers. These three systems are capable of performing collision-induced dissociation to fragment peptides, allowing the intact peptide, as well as the fragmented ions, to be monitored. Many companies deploy these systems for drug-level testing in drug development and in clinical trial settings or for small-molecule biomarker assays, such as for testosterone and vitamin D. These platforms serve multiple purposes and are well established in laboratories. Both the ion trap and the quadrupole ion trap systems have become more common in recent years, utilized in research for developing protein biomarkers. The cost of these MS configurations is reasonable or is becoming competitive with other platforms in the clinical setting.

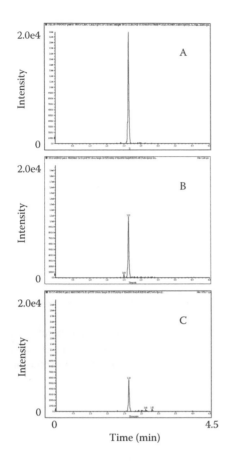

[pE]EDI⌈PE⌈V⌉SCIHNGLR
        y11 y9 y8

Trace	Q1 (m/z)	Q1 Ion	Q3 (m/z)	Q3 Ion
A	869.5	M+2H	1269.7	y"11+
B	869.5	M+2H	944.5	y"8+
C	869.5	M+2H	1043.6	y"9+

FIGURE 2.2 The three transitions monitored for the quantitation of rat P1NP. Three transitions of the doubly charged ion of a P1NP tryptic peptide are monitored in rat serum for quantitation of the protein to ensure specificity and selectivity. Signal intensity does vary between the three transitions, and they are averaged for quantitation purposes.

Higher-end MS systems, such as the quadrupole time-of-flight (QTOF), Orbitrap, or Fourier transform-ion cyclic resonance mass spectrometer (FT-ICR), are commonly used in proteomics experiments due to their higher resolution and mass accuracy. These systems in particular are often thought of as difficult to use for quantitation experiments due to design issues. For example, in QTOF mass spectrometers, subtle temperature changes in the laboratory could result in peak drift in the mass spectra. Manufacturers have implemented sophisticated designs and expensive electronics to overcome these problems so that the systems are more robust, resulting in more researchers using these for quantitation. The challenge at this time is that these systems tend to cost two to three times more than triple-quadrupole mass spectrometers, require more maintenance, and often need highly trained personnel.

Several factors should be considered with the selection of the MS platform. One of the most important considerations for protein biomarker assays is the mass range of the system. Although most systems have a mass range suitable for the detection of most peptides, detection of some larger peptides with the best ion in the range of 1500–2000 may be an issue. Other multiply charged ions may have a strong enough signal for the assay needs or may be in a range that is noisy. Therefore, systems with a larger mass range are beneficial for broad application.

Matrix-assisted laser desorption/ionization (MALDI) is another soft ionization technique used in MS for the analysis of biomolecules, but it has the advantage of producing fewer multiply charged ions than in ESI. The difficulty with interfacing MALDI with chromatographic systems is that MALDI is a solid-phase technique, so the eluent must be mixed with matrix and dried into a crystal prior to analysis. Finally, MALDI analysis has limited quantitation capabilities and therefore is better utilized in discovery-based proteomics.

2.3.2 OTHER INTERFACES AND DETECTORS

Although mass spectrometers are currently the most widely used type of detector for quantitative protein biomarker assays that employ chromatography, other detectors have some advantages that may make them more attractive, depending on the needs of the assay. For example, HPLC-evaporative light-scattering detection (ELSD) was used to develop a quantitative assay for surfactant protein B, which is related to respiratory function and lung injury [44]. The HPLC assay was found to be more efficient than available ELISA techniques due to the need to remove lipids in the sample for the ELISAs. The challenge with most of these detectors is the interference from the components in the biological matrix or distinguishing analytes in such a complex mixture. Detectors, such as diode arrays, ELSDs, and fluorescent detectors, are directly interfaced to chromatographic systems and are often used in the analysis of proteins and biological samples. However, these detectors are not selected as often for quantitative protein biomarker assays in comparison to mass spectrometers, and this is illustrated by the list of protein biomarkers in Table 2.1.

Ultraviolet (UV) detection is one of the simplest interfaces to implement with HPLC, yet it presents some challenges in regard to protein and peptide quantitation in complex biological samples. UV detects proteins and peptides at relatively high concentrations, with a 10^{-6} M detection range. However, the dynamic range for UV detection is limited, so the constituents in complex biological samples have to be fractionated effectively for UV detection to work. A further complication is that the response factor differs from protein to protein or peptide to peptide. Typically, at a wavelength of 214 nm, the carbon nitrogen peptide bond is monitored. The absorption maximum of all amino acids is below 210 nm, but this wavelength is not the most sensitive due to interferences from solvents and other constituents. The wavelength of 280 nm can be employed for the detection of aromatic amino acid residues (tyrosine, tryptophan, histidine, and phenylalanine), but not all peptides contain these residues. UV detection has limited selectivity among proteins and peptides, so getting the peak of interest isolated in a complex sample without other contaminants underneath the peak can be challenging. Due to its limited selectively, UV detection is often limited to determining total protein/peptide concentration or the identification of fractions of interest. Diode arrays provide the ability to detect a range over the UV-visible spectrum, yet have mostly the same limitations.

The fluorescent detector may be more preferable to UV detection since it provides a high level of sensitivity and some selectivity. Fluorescence and UV laser-induced fluorescence (UV-LIF) detection are approximately 100 times more sensitive than traditional UV detection. Proteins and peptides can be fluorescently tagged with reagents like fluorescein isothiocyanate or derivatized for detection, and even the native fluorescence of certain amino acids, such as tyrosine, can be capitalized on [45,46]. One example of a biomarker assay that utilizes fluorescent detection is the assay for pentosidine, an advanced glycation end product that correlates with the presence and severity of diabetic complications in patients with type II diabetes. Pentosidine is a fluorescent cross-link between arginine and lysine, and its fluorescent properties make it easy to detect by HPLC with fluorescent detection [47].

Nuclear magnetic resonance (NMR) is a technique often used in metabonomics experiments for the profiling and quantitation of metabolites in biological fluids. Profiles of metabolites are being used to assess the risk for disease and even diagnosis [48,49]. Although these platforms are often applied to small molecules, application to peptides is possible. The NMR can be interfaced to chromatographic systems and partnered with MS. NMR does not offer the same sensitivity as MS yet does not have the issues in regard to the ionization efficiency of analytes that the mass spectrometer has. The challenge comes with the interference of background molecules that complicate spectra. Steps can be taken to overcome this, such as labeling and derivatization, yet biological samples have a high number of interferences that may be preventive [50].

2.3.3 DATA SYSTEMS

Another consideration for platform selection is the ability to interface to data repositories for data analysis, reporting, and storage. Most instrument manufacturers provide software and computer systems to operate the instrumentation and process the raw data. Once the final protein or peptide measurement has been determined, the data may need to be transferred to an external server with database programs to integrate the data with other information. Since data files on most systems can be exported in a variety of file formats, and most data systems will take a variety of types of data files, interfacing between systems is typically not an issue. Protein biomarker data can be recorded with information on the assay, along with demographic data on the sample or patient and with values of other endpoints or biomarkers. Additional information, such as age, disease state, and time of collection, aids in the review and confirmation of data for the performance of the biomarker and the assay. Many data systems in hospital and clinical settings have this ability, and laboratory analyzers are directly interfaced to the data systems, enabling real-time data collection into the database.

Consideration for data storage and format with biomarkers is more about the ability to interrogate the data. Having a system that can collect the end result with the sample demographics and other endpoints is powerful for the utility of the biomarker, such as understanding the reference range of a particular biomarker within a certain patient population. An in-depth database allows for retrospective analysis of data sets to investigate new hypotheses and define performance of the biomarkers, enabling better interpretation of results in future analyses. Many are using these types of data-bases to provide retrospective studies when prospective studies are not possible. For example, hospital databases were used to examine the relationship between serum potassium levels and in-hospital mortality in acute myocardial infarction patients now that different therapies are used, resulting in new recommendations for serum potassium levels in these patients [51]. Another retrospective study queried hospital databases to validate disease classification codes for acute kidney injury with serum creatinine levels that helped understand the diagnostic performance and limitations of the code [52]. Having the data stored in such a manner makes interrogation of the data easier and improves overall quality and use of these biomarkers.

2.4   TECHNIQUES FOR SAMPLE PREPARATION AND HANDLING

The complexity of the sample preparation for a quantitative protein assay depends on the sensitivity needed, the required robustness, time available to conduct the assay, and the tools available to the researcher. Choosing the best sample-handling and preparation techniques is important because these factors often dictate how robust, selective, and sensitive an assay will be, contributing to the overall success of the assay. Many obvious factors should be considered in the development of a protein assay. Being aware of protein solubility is important, and knowledge of how to keep the protein soluble without interfering with the assay is key. For example, detergents are used to maintain protein solubility yet can overwhelm LC-MS systems due to their concentration and their potential to interfere with ionization. Beginning conditions for the injection onto the LC-MS system are also important. Various techniques for cleaning up the sample may leave the sample in a solution that is not compatible with the starting conditions of the chromatography.

This section provides a brief overview of some of the most common sample preparation techniques used for protein quantitative assays. Since the majority of chromatographic-based assays for protein biomarkers are LC-MS assays, most of the discussion focuses on the workflows for quantitative analyses of proteins in bio-logical samples that employ enzymatic digestion to produce peptides for quantita-tion. Most protein biomarker assays follow one of the three experimental schemes outlined in Figure 2.3, with all workflows utilizing enzymatic digestion, followed at some point by LC-MS analysis. Immunoprecipitation techniques aid in the reduction of sample complexity, providing gains in the sensitivity of the assay. These proce-dures are most commonly used, yet additional steps may be needed for the devel-opment of assays to detect low-abundant proteins. These workflows are merely the backbone of the assay, with many additional possibilities.

2.4.1   PROTEIN DIGESTION

Due to the complex nature of biological samples, intact proteins are rarely the target for detection in protein biomarker assays. Usually, proteins are broken down into smaller, more manageable peptides by enzymatic digestion. Enzymes that cleave at known, well-defined sites are preferable, as one can predict the peptides that will be formed and design the assay around these predicted peptides. The most

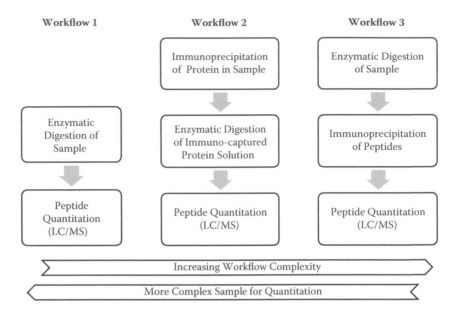

FIGURE 2.3 Basic experimental design for protein biomarker assays by LC-MS. Three basic workflows exist for quantitative protein biomarker assays by LC-MS. The process chosen depends on the sensitivity needed for the assay and on reagent availability. The addition of an internal standard occurs at an appropriate point in the process, and this internal standard is typically the peptide being monitored with the addition of an isotopic label. Additional sample preparation may be needed prior to these workflows.

common enzyme used is trypsin, which specifically cleaves at the C-terminal side of lysines and arginines in the protein sequence. This enzyme is available at very high purities with good activity and cleaves proteins at predictable sites. Lys-C and chymotrypsin are examples of other specific enzymes. Nonspecific enzymes (proteases) or acid hydrolysis can also be used to break down the protein and are usually used in the analysis of adducts or modifications to the protein that relate to disease state. For example, the assay for pentosidine, a biomarker example discussed previously, requires acid hydrolysis of the plasma sample to recover the pentosidine before quantitative analysis [47]. Some protein biomarker assays are for peptides of the protein that are formed from enzymatic cleavages in vivo. For example, the breakdown of collagen type II occurs in patients with osteoarthritis and rheumatoid arthritis, and monitoring urine for the peptide fragments of collagen type II produced by matrix metalloproteinase activity may be useful for early diagnosis [53].

During the digestion procedure, certain steps are taken to maintain the reproducibility of the digestion. Often, steps to optimize the digestion time and temperature used for digestion are conducted. Some assays utilize the addition of an internal standard peptide to monitor digestion efficiency [31]. These peptides are often a longer version of the peptide being used for quantitation that includes an additional cleavage site for the enzyme. Almost every assay that utilizes enzymatic digestion also includes a reduction and alkylation step to reduce any disulfide bonds in the protein and prevent them from re-forming after digestion. This step results in cysteine residues having an adduct, increasing the mass of any cysteine-containing peptides.

2.4.2 *In Silico* Analyses

In silico analyses are one of the most powerful, and sometimes most underutilized, tools in assay development. Utilizing these types of tools can provide knowledge of physical chemical properties that aid in the selection of sample-handling procedures, chromatographic conditions, and MS transitions. Novel and newly discovered proteins pose some challenges in that limited information may be available. In some cases, only a genomics sequence may be known, and the protein sequence is derived from there. Yet, having only the sequence does not mean these tools cannot be useful. Once peptides are selected, the sample-handling techniques can be chosen based on the sensitivity needed for the assay, the physical chemical properties of the peptide, and the technique that best fits the needs of the assay.

The most common type of *in silico* analysis is the digestion of the protein to determine what peptides are formed and which would make good candidates for quantitation. The list of peptides formed from the theoretical enzymatic digestion enables a starting list of transitions for the mass spectrometer to investigate. Others have taken the enzymatically digested protein, generated MS data, and then performed a database search to match ions to protein-specific peptides, which was the methodology used to develop the assay for ceruloplasmin (CP) and provides valuable experimental information in conjunction with the *in silico* digestion [25].

Certain characteristics of the peptides make them more amenable for quantitation than others. For example, the size of the peptide is important. Very small peptides that consist of only a few amino acids can be difficult to detect as they elute early in a reversed-phase separation with many other small peptides. These peptides may not be very specific since they may be formed by many other proteins in the mixture. Larger peptides pose difficulties because they are late eluters and difficult to fragment in the mass spectrometer.

Another example is potential complication of the analysis with sites of post-translational modifications on the peptides. Certain sites and sequences are known for their potential to be modified. Cysteines can form disulfide bonds; methionines are often oxidized; serines and threonines can be sites of O-linked glycosylation; and the sequence N-X-S/T indicates a potential site for N-linked glycosylation. If the modification is conserved, then the mass of the peptide is constant, which is not the case if the post-translational modification is of low abundance or heterogeneous.

These considerations are important when evaluating the potential peptides to use for quantitation, yet the best candidates for quantitation ultimately need to be determined experimentally. How the peptides actually respond in the chromatographic and MS systems (elution time and ionization efficiency) is important in selection. Table 2.2 illustrates the *in silico* trypsin digestion of the rat N-terminal propeptide of alpha-1 type 1 collagen that makes up a portion of P1NP. Peptides were chosen for further investigation based on properties such as their size and potential for modification.

TABLE 2.2

In Silico Trypsin Digestion of the Rat N-Terminal Propeptide of Alpha-1
Type 1 Collagen (P02454)

Peptide	Amino Acid Residues	Tryptic Peptides	Molecular Mass (m/z)	Observations
1	23–37	QEDIPEVSCIHNGLR	1692.8	Gln to pyroGlu
2	38–67	VPNGETWKPDVCLICI CHNGTAVCDGVLCK	3187.5	Too long and potential N-linked glycosylation site
3	68–77	EDLDCPNPQK	1158.5	Potential incomplete cleavage after residue 77
4	78	R	175.2	Potential incomplete cleavage after residue 77
5	79–104	EGECCPFCPEEYVS PDAEVIGVEGPK	2783.2	Too large and contains multiple Cys
6	105–113	GDPGPQGPR	880.4	Not unique
7	114–151	GPVGPPGQDGIPGQPGLPGPP GPPGPPGPLGLGGNFAS	3347.7	Too large

In silico analyses should also be performed to ensure the peptide is unique in the known proteome. If the peptide sequence is conserved in a family of proteins, it would not be a good candidate for selection since it would not have a high level of specificity. However, in some cases a unique peptide may not be possible to identify in those families, so the readout of that signal has the caveat of representing a group of proteins. If the marker changes with the disease state in such a way that that level of specificity is not needed, then that peptide is sufficient. The best-case scenario is that the peptides used for quantitation are unique to the target protein.

The *in silico* digestion of proteins provides insight in determining the peptides to be used for quantitation and for the design of the assay. The previous example illustrates the selection of the peptide for quantitation in the P1NP assay in rat serum. The N-terminal propeptide of alpha-1 type 1 collagen, a portion of P1NP, consists of residues 23–151. An *in silico* trypsin digestion was performed, followed by basic local alignment search tool (BLAST) searches to determine the uniqueness of the peptides in the rat proteome. Peptide 1 was the best choice due to size, potential modification sites, and unambiguous identification of P1NP [31]. The glutamine at the N-terminal is modified to a pyroglutamate, yet this modification is conserved.

2.4.3 IMMUNOPRECIPITATION TECHNIQUES

When antibodies are available, immunoprecipitation techniques provide a powerful sample fractionation process, often needed for sensitivity gain when analyzing proteins and peptides at low detection levels. Antibodies to the protein of interest or to peptides from the digestion can be used to extract the analytes of interest from other components in the biological sample. A high affinity or specificity to the protein or peptide is not required, as is needed for ELISAs, because this experimental step is performed as part of sample cleanup. This technique is often performed off line, such that antibodies are immobilized, often to a plate or beads, and incubated with the sample to allow binding to occur. The protein is then eluted off the antibody and then prepared for downstream analysis, which may include enzymatic digestion. Immunoprecipitation is typically utilized when low levels of a protein need to be detected, as in the case for the CP assay. Physiological levels of CP are very low in newborns, so screening for Wilson's disease requires an assay with high sensitivity and precision [25]. In some cases, the elution step is skipped, and the protein is digested while bound to the antibody, resulting in the formation of peptides from both the protein and the antibody. In some cases, the proteins can be difficult to elute efficiently, so digesting them directly on the antibody may be preferable. The assay for myocin light chain 1 (Myl l/3) utilizes on-bead digestion since it facilitates higher recovery of the peptide, without interference from peptides generated from the antibody, and speeds the sample preparation process [27].

Utilization of immunoprecipitation techniques does provide some challenges. If commercial antibodies are not available, the generation of antibodies can take months. A viable system must be identified in which the antibody can be developed; the antibody must be isolated and then screened to test the selectivity for the protein or peptide. Many commercial companies produce reliable and reproducible reagents, and catalogs contain hundreds of antibodies for use in multiple applications. Antibodies developed for other techniques, such as for immunohistochemistry, need to be tested for their use with immunoprecipitation as not all antibodies may work in this application. Many individual research laboratories produce their own antibody reagents for use in assays. The difficulty with these antibodies is that they are not widely available to others, and other laboratories may not be able to produce an antibody with similar binding characteristics, contributing to the inability to reproduce results between laboratories. Sometimes, the cost of making the antibody can be prohibitive since the steps from start to finish may be labor intensive. Often, polyclonal antibodies are developed for these procedures since these antibodies take less time to develop than monoclonals. Other challenges include determining the best binding and elution conditions for the immunoprecipitation process, as well as ensuring that the buffers used are compatible between experimental steps, especially once the sample is to be injected onto the chromatographic system.

2.4.4 Isotopic Labeling of Internal Standards

Internal standards that are isotopically labeled aid in many aspects of LC-MS assay development and performance. Most commonly, synthetic peptides identical to the peptide of interest are labeled with a stable isotope, such as carbon-13 or nitrogen-15, on the lysine or arginine residue at the end of the tryptic peptide. The isotopically labeled peptides elute at the same time as the naive peptide and fragment in the same manner. Deuterium-labeled peptides are not selected due to the fact that the isotope will actually delay the elution. The mass spectrometer can detect the mass difference between the naïve and isotopically labeled peptides. Most of the biomarkers listed in Table 2.1 rely on the use of isotopically stable-labeled peptides as internal standards, demonstrating the popularity of this technique.

For the detection of low-abundance proteins in biological fluids, researchers have employed an approach termed stable isotope standard capture with antipeptide antibodies (SISCAPA) [54]. SISCAPA is a combined assay using signature peptides and their associated stable isotope-labeled internal standards that are enriched from sample digests by antipeptide antibodies, followed by targeted LC-MS readouts. Briefly, a researcher synthesizes a standard for a signature peptide. The synthetic stable isotope peptide is then used to generate a highly specific antipeptide antibody, utilized to enrich the signature peptide in a complex mixture. The antipeptide antibody is then immobilized on magnetic beds, and a protein digest spiked with the stable isotope peptide is subjected to affinity chromatography. After affinity enrichment, both the native peptide and stable isotope-labeled peptide are eluted and analyzed via MS. These techniques have been used to develop assays for troponin I and interleukin (IL) 33, a benefit for the analysis of IL-33 in that no commercial ELISA assay existed [26].

2.5 CONSIDERATIONS FOR ASSAY DEPLOYMENT

Biomarker assays are deployed in many different laboratory settings, from a research laboratory to the clinical laboratory, and the needs and requirements of each are very different. Assays in the clinical laboratory often need to have simple and robust sample preparation procedures, exist on standard platforms, be standardized across laboratories, and have the capabilities for high throughput. The setting in the research laboratory may look very different in that assays may have more complicated sample-handling procedures and instrument configurations, and little need may exist for standardization between laboratories. Three main factors shape the requirements of the assay: the extent of assay deployment, the instrumentation used, and the throughput requirements.

2.5.1 BIOMARKER DEPLOYMENT

A major consideration that is often overlooked during the development of an assay is to what extent other laboratories will use the assay. Some biomarker assays are considered proprietary, in that the company does not want others to know the endpoint being measured, the assay being used, or how it is being applied. For example, proprietary assays that demonstrate the effectiveness of a drug can provide the pharmaceutical company an advantage over competitors, so the assay will only be deployed within the company. In other instances, the assay may need to be used by multiple laboratories. For example, pharmaceutical companies want to use widely accepted biomarkers to ensure the safety of a drug when filing a drug application to ensure the acceptance of that biomarker and data by regulatory agencies. When all companies are using the same assay, the regulatory agencies have more confidence in the data since the platform for analysis can be recommended and guidelines for data set.

Knowing how the biomarker is going to be deployed can shape many factors during assay development. More complicated assays are more difficult to transfer and affect the reproducibility between laboratories. Some assays can be so complicated that transfer to other laboratories is difficult, often resulting in the external laboratory making modifications to the assay or not being able to reproduce the results. Multisite comparisons of protein measurements in plasma by LC-MS have been conducted and illustrate the robustness of these assays when transferring them between laboratories, as well as when deploying the assay on similar platforms from varying manufacturers [55]. Any assay that is to be transferred must be robust in all aspects to be successful.

2.5.2   Instrumentation

The instrument platform choice is important if the assay is going to be utilized across many laboratories. The cost to purchase, install, and maintain the equipment is part of the consideration, and if the equipment is already present in the laboratory, it is more likely the assay can be successfully deployed. For example, contract laboratories offer assay services on certain platforms and most likely would not purchase new equipment for a special request. Also, many laboratories do not have the resources to maintain higher-end systems that are costly. Triple-quadrupole mass spectrometers interfaced to conventional HPLC systems are commonly found in contract laboratories offering bioanalytical services for determining drug concentration in blood samples. Without modification to the instrument configuration, these systems can be deployed for quantitative protein assays. Currently, nanoflow chromatography systems are rarely found in contract laboratories outside those for proteomic discovery services. These systems require trained personnel, and the connections with electrospray sources are delicate, so assays that have been developed on nanoflow systems often need to be redeveloped on other chromatographic systems to be widely disseminated. However, many laboratories are beginning to utilize microflow systems, so in this case finding the right laboratory partner to deploy the assay would be easier, and any assay on a conventional HPLC system would be easily deployed at most laboratories with the technology.

2.5.3   Sample Throughput

The final consideration for assay deployment is the sample throughput requirements. If the assay is going to be deployed to thousands of samples, such as those from clinical trials, time becomes one of the most critical factors. Typically, the two areas that affect the overall time for an assay are the sample preparation and instrument run time. The sample preparation for a protein biomarker assay can range from a simple one-step procedure to steps that require overnight digestions and incubations. A sample preparation that requires several hours to complete or occurs over multiple days may not be ideal for a large number of samples. On the other hand, if the assay is to be deployed to a small number of samples, more complicated handling and procedures may be acceptable. Most labs have some level of automation, from 96-well plate technologies to robotics capable of handling thousands of samples, and these systems have to be configured specifically for the assay. Most methods can be highly automated, yet automation can be more difficult and tedious for proteins at low levels.

As described, instrument run time also dictates sample throughput. The time needed to effectively separate the analytes is typically the rate-limiting step, so utilizing UPLC systems can drastically reduce the run time of the assay [40]. For example, the HPLC assay for erythropoietin, a glycoprotein that controls red blood cell production, requires a 20-min run time, yet the UPLC only requires 4 min [56]. When applied to 100 samples, the difference in run time for all samples is 1600 min, or just over 1 day, providing a significant time savings. Most assays will transfer to UPLC systems yet require that the laboratory have access to this type of LC system.

Other methods can be considered to reduce run times that involve creative use of the chromatographic system. For example, stacked injections can be performed, yet they are difficult for the analysis of complex samples in that the peak of interest would need to be separated from any other co-eluters in subsequent injections. This technique would require additional chromatographic systems and multiple columns. With multiple pumps, systems utilizing two-dimensional chromatography can be programmed in such a manner that total run times can be reduced. As an analyte is being bound in the first dimension, another sample can be eluting in the second. An evaluation of the time savings, as compared to the time to run and maintain a second system, would need to be conducted to determine if worth the effort.

If multiple biomarkers are to be deployed in a laboratory, certain panels of biomarkers can be multiplexed together so they are analyzed in a single run. Analytes with similar properties make prime candidates for multiplexing, especially if these markers are typically run for the same disease states. When compared to a single assay, the total run time may be longer due to the need to separate multiple analytes, yet the overall time savings to run and maintain one assay versus two may be beneficial. Many examples of multiplexing exist for targeted proteomic experiments and profiling small molecule biomarkers, and these multiplexing techniques are beginning to emerge for quantitative protein applications [18,57–59]. For example, four clinically validated cardiovascular biomarkers (lactate dehydrogenase B (LDH-B), creatine kinase MB heterodimer, myoglobin, and troponin I) have been multiplexed into a single assay to speed and improve biomarker evaluation [29].

2.6 FUTURE STATE

Development and deployment of protein biomarker assays on chromatographic and MS platforms is becoming more common in individual laboratories [60]. However, these assays have not yet become routine across laboratories, in laboratory hospitals, or in contract laboratories [10]. In recent years, we have seen small-molecule biomarker assays take off within these settings. LC-MS assays for testosterone, estradiol, vitamin D analogs, and T_3/T_4 (triiodothyronine/thyroxine) are standard assays in the clinical setting, creating a familiarity with the platform that did not exist previously. LC-MS is even considered the gold standard in some cases, replacing antibody-based assays. Advances in instrumentation will continue to influence and improve these assays, as well as better interfaces for the analyst who is not an expert in chromatography or MS. Currently, many of the quantitative protein biomarker assays are being used for the cross-validation of ELISAs or for prescreening antibodies for immunoassay development [23,27]. ELISAs are typically more cost effective for routine sample analysis yet can take more time to develop. As the demand for more sensitive and selective protein biomarker assays increases, more chromatography-based assays will begin emerging as routine clinical assays.

ACKNOWLEDGMENTS

Thank you to Christopher Colangelo, Carol Fritz, and Kimberly Navetta for their insights on the content of this chapter and contributions to figure contents.

REFERENCES

1. N. Rifai, M. A. Gillette, and S. A. Carr, Protein biomarker discovery and validation: the long and uncertain path to clinical utility, *Nature Biotechnology* 24 (2006) 971–983.
2. T. Niwa, Biomarker discovery for kidney diseases by mass spectrometry, *Journal of Chromatography B* 870 (2008) 148–153.
3. J. Ozer et al., Enhancing the utility of alanine aminotransferase as a reference standard biomarker for drug-induced liver injury, *Regulatory Toxicology and Pharmacology* 56 (2010) 237–246.
4. H. Shui, T. Huang, S. Ka, P. Chen, Y. Lin, and A. Chen, Urinary proteome and potential biomarkers associated with serial pathogenesis steps of focal segmental glomeruloscle-rosis, *Nephrology Dialysis Transplantation* 23 (2008) 176–185.
5. S. A. Varghese, T. B. Powell, M. N. Budisavljevic, J. C. Oates, J. R. Raymond, J. S. Almeida, and J. M. Arthur, Urine biomarkers predict the cause of glomerular disease, *Journal of the American Society of Nephrology* 12 (2007) 913–922.
6. N. Perico, A. Benigni, and G. Remuzzi, Present and future drug treatments for chronic kidney diseases: evolving targets in renoprotection, *Nature Reviews* 7 (2008) 936–953.
7. J. V. Bonventre, V. S. Vaidya, R. Schmouder, P. Feig, and F. Dieterle, Next-generation biomarkers for detecting kidney toxicity, *Nature Biotechnology* 28 (2010) 436–440.
8. J. Ozer, M. Ratner, M. Shaw, W. Bailey, and S. Schomaker, The current state of serum biomarkers of hepatotoxicity, *Toxicology* 245 (2008) 194–205.
9. S. Schomaker, R. Warner, J. Bock, K. Johnson, D. Potter, J. Van Winkle, and J. Aubrecht, Assessment of emerging biomarkers of liver injury in human subjects, *Toxicological Sciences* (2013) Epub ahead of print.
10. G. Paulovich, J. R. Whiteaker, A. N. Hoofnagle, and P. Wang, The interface between biomarker discovery and clinical validation: the tar pit of the protein biomarker pipeline, *Proteomics—Clinical Applications* 2 (2008) 1386–1402.
11. D. Amacher, The discovery and development of proteomic safety biomarkers for the detection of drug-induced liver toxicity, *Toxicology and Applied Pharmacology* 245 (2010) 134–142.
12. L. Anderson, Candidate-based proteomics in the search for biomarkers of cardiovascu-lar disease, *Journal of Physiology* 563.1 (2005) 23–60.
13. C. Fenselau, A review of quantitative methods for proteomic studies, *Journal of Chromatography* B 855 (2007) 14–20.
14. J. A. Jones, L. Kaphalia, M. Treinen-Moslen, and D. C. Liebler, Proteomic char-acterization of metabolites, protein adducts, and biliary proteins in rats exposed to 1,1-dichloroethylene or diclofenac, *Chemical Research in Toxicology* 16 (2003) 1306–1317.
15. Q. Wu, H. Yuan, L. Zhang, and Y. Zhang, Recent advances on multidimensional liq-uid chromatography-mass spectrometry for proteomics: from qualitative to quantitative analysis—a review, *Analytica Chimica Acta* 731 (2012) 1–10.
16. P. Khatri, M. Sirota, and A. J. Butte, Ten years of pathway analysis: current approaches and outstanding challenges, *PLoS Computational Biology* 8 (2012) e1002375.
17. V. Lange, P. Picotti, B. Domon, and R. Aebersold, Selected reaction monitoring for quantitative proteomics: a tutorial, *Molecular Systems Biology* 4 (2008) 222.

18. K. Yocum and A. M. Chinnaiyan, Current affairs in quantitative targeted proteomics: multiple reaction monitoring-mass spectrometry, *Briefings in Functional Genomics and Proteomics* 8 (2009) 145–157.
19. S. Makawita, and E. P. Diamandis, The bottleneck in the cancer biomarker pipeline and protein quantification through mass spectrometry-based approaches: current strategies for candidate verification, *Clinical Chemistry* 56 (2010) 212–222.
20. S. A. Carr and L. Anderson, Protein quantitation through targeted mass spectrometry: the way out of biomarker purgatory? *Clinical Chemistry* 54 (2008) 1749–1752.
21. B. Han and R. E. Higgs, Proteomics: from hypothesis to quantitative assay on a single platform. Guidelines for developing MRM assays using ion trap mass spectrometers, *Briefings in Functional Genomics and Proteomics* 7 (2008) 340–354.
22. C. Hunter and L. Basa, "Detecting a Peptide Biomarker for Hypertension in Plasma: Peptide Quantitation Using the 4000 Q TRAP™ System." Applied Biosystems Proteomics Technology Note.
23. E. Kuhn, J. Wu, J. Karl, H. Liao, W. Zolg, and B. Guild, Quantification of C-reactive protein in the serum of patients with rheumatoid arthritis using multiple reaction monitoring mass spectrometry and ^{13}C-labeled peptide standards, *Proteomics* 4 (2004) 1175–1186.
24. D. K. Williams, Jr., and D. C. Muddiman, Absolute quantification of C-reactive protein in human plasma derived from patients with epithelial ovarian cancer utilizing protein cleavage isotope dilution mass spectrometry, *Journal of Proteome Research* 8 (2009) 1085–1090.
25. A. deWilde, K. Sadilkova, M. Sadilek, V. Vasta, and S. H. Hahn, Tryptic peptide analysis of ceruloplasmin in dried blood spots using liquid chromatography-tandem mass spectrometry: application to newborn screening, *Clinical Chemistry* 54 (2008) 1961–1968.
26. Eric Kuhn et al., Developing multiplexed assays for troponin i and interleukin-33 in plasma by peptide immunoaffinity enrichment and targeted mass spectrometry, *Clinical Chemistry* 55 (2009) 1108–1117.
27. M. J. Berna, Y. Zhen, D. E. Watson, J. E. Hale, and B. L. Ackermann, Strategic use of immunoprecipitation and LC/MS/MS for trace-level protein quantification: myosin light chain 1, a biomarker of cardiac necrosis, *Analytical Chemistry* 79 (2007) 4199–4205.
28. M. Berna and B. Ackermann, Increased throughput for low-abundance protein biomarker verification by liquid chromatography/tandem mass spectrometry, *Analytical Chemistry* 81 (2009) 3950–3956.
29. C. Huillet et al., Accurate quantification of cardiovascular biomarkers in serum using protein standard absolute quantification (PSAQ™) and selected reaction monitoring, *Molecular & Cellular Proteomics* 11 (2012) M111.008235.
30. M. Berna, L. Ott, S. Engle, D. Watson, P. Solter, and B. Ackermann, Quantification of NTproBNP in rat serum using immunoprecipitation and LC/MS/MS: a biomarker of drug-induced cardiac hypertrophy, *Analytical Chemistry* 80 (2008) 561–566.
31. B. Han et al., Development of a highly sensitive, high-throughput, mass spectrometry-based assay for rat procollagen type-I N-terminal propeptide (P1NP) to measure bone formation activity, *Journal of Proteome Research* 6 (2007) 4218–4229.
32. D. R. Barnidge, M. K. Goodmanson, G. G. Klee, and D. C. Muddiman, Absolute quantification of the model biomarker prostate-specific antigen in serum by LC-MS/MS using protein cleavage and isotope dilution mass spectrometry, *Journal of Proteome Research* 3 (2004) 644–652.
33. S. Bolisetty and A. Agarwal, Urine albumin as a biomarker in acute kidney injury, *American Journal of Physiology: Renal Physiology* 300 (2011) F626–F627.
34. M. J. Edelmann, Strong cation exchange chromatography in analysis of posttranslational modifications: innovations and perspectives, *Journal of Biomedicine and Biotechnology* 2011 (2011) 10.1155/2011/936508.

35. D. S. Hage, Affinity chromatography: a review of clinical applications, *Clinical Chemistry* 45 (1999) 593–615.
36. R. Kange, U. Selditz, M. Granberg, U. Lindberg, G. Ekstrand, B. Ek, and M. Gustafsson, Comparison of different IMAC techniques used for enrichment of phosphorylated peptides, *Journal of Biomolecular Techniques* 16 (2005) 91–103.
37. G. B. Irvine, Size-exclusion high-performance liquid chromatography of peptides: a review, *Analytica Chimica Acta* 352 (1997) 387–397.
38. G. J. Opiteck, S. M. Ramirez, J. W. Jorgenson, and M. A. Moseley III, Comprehensive two-dimensional high-performance liquid chromatography for the isolation of overexpressed proteins and proteome mapping, *Analytical Biochemistry* 258 (1998) 349–361.
39. S. D. Garbis, T. I. Roumeliotis, S. I. Tyritzis, K. M. Zorpas, K. Pavlakis, and C. A. Constantinides, A novel multidimensional protein identification technology approach combining protein size exclusion prefractionation, peptide zwitterion-ion hydrophilic interaction chromatography, and nano-ultraperformance RP chromatography/nESI-MS^2 for the in-depth analysis of the serum proteome and phosphoproteome: application to clinical sera derived from humans with benign prostate hyperplasia, *Analytical Chemistry* 83 (2011) 708–718.
40. M. E. Swartz, UPLC: an introduction and review, *Journal of Liquid Chromatography & Related Technologies* 28 (2005) 1253–1263.
41. L. Novakova and H. Vlckova, A review of current trends and advances in modern bioanalytical methods: chromatography and sample preparation, *Analytica Chimica Acta* 656 (2009) 8–35.
42. I. Francois, K. Sandra, and P. Sandra, Comprehensive liquid chromatography: fundamental aspects and practical considerations – A review, *Analytica Chimica Acta* 641 (2009) 14–31.
43. L. Anderson and C. Hunter, Quantitative mass spectrometric multiple reaction monitoring assays for major plasma proteins, *Molecular and Cellular Proteomics* 5 (2006) 573–588.
44. C. Paschen and M. Griese, Quantitation of surfactant protein B by HPLC in bronchoalveolar lavage fluid, *Journal of Chromatography B* 814 (2005) 325–330.
45. K. C. Chan, T. D. Veenstra, and H. J. Issaq, Comparison of fluorescence, laser-induced fluorescence, and ultraviolet absorbance detection for measuring HPLC fractionated protein/peptide mixtures, *Analytical Chemistry* 83 (2011) 2394–2396.
46. S. Saraswat, B. Snyder, and D. Isailovic, Quantification of HPLC-separated peptides and proteins by spectrofluorimetric detection of native fluorescence and mass spectrometry, *Journal of Chromatography B* 902 (2012) 70–77.
47. J. L. Scheijen, M. P. H. van de Waarenburg, C. D. A. Stehouwer, C. G. Schalkwijk, Measurement of pentosidine in human plasma protein by a single-column high-performance liquid chromatography method with fluorescence detection, *Journal of Chromatography B* 877 (2009) 610–614.
48. M. G. Barderas, C. M. Laborde, M. Posada, F. de la Cuesta, I. Zubiri, F. Vivanco, and G. Alvarez-Llamas, Metabolomic profiling for identification of novel potential biomarkers in cardiovascular diseases, *Journal of Biomedicine and Biotechnology* 2011:790132. doi: 10.1155/2011/790132
49. E. Garcia, C. Andrews, J. Hua, H. L. Kim, D. K. Sukumaran, T. Szyperski, and K. Odunsi, Diagnosis of early stage ovarian cancer by 1H NMR metabonomics of serum explored by use of a microflow NMR probe, *Journal of Proteome Research* 10 (2011) 1765–1771.
50. S. Zhang, G. A. N. Gowda, T. Ye, and D. Raftery, Advances in NMR-based biofluid analysis and metabolite profiling, *Analyst* 135 (2010) 1490–1498.
51. A. Goyal, J. A. Spertus, K. Gosch, L. Venkitachalam, P. G. Jones, G. Van den Berghe, and M. Kosiborod, Serum potassium levels and mortality in acute myocardial infarction, *The Journal of the American Medical Association* 307 (2012) 157–164.

52. Y. J. Hwang, S. Z. Shariff, S. Gandhi, et al, Validity of the *International Classification of Diseases, Tenth Revision* code for acute kidney injury in elderly patients at presentation to the emergency department and at hospital admission, *BMJ Open* 2 (2012) e001821.

53. O. V. Nemirovskiy, D. R. Dufield, T. Sunyer, P. Aggarwal, D. J. Welsch, and W. R. Mathews, Discovery and development of a type II collagen neoepitope (TIINE) biomarker for matrix metalloproteinase activity: from in vitro to in vivo, *Analytical Biochemistry* 361 (2007) 93–101.

54. J. R. Whiteaker, L. Zhao, H. Y. Zhang, L. C. Feng, B. D. Piening, L. Anderson, and A. G. Paulovich, Antibody-based enrichment of peptides on magnetic beads for mass-spectrometry-based quantification of serum biomarkers, *Analytical Biochemistry* 362 (2007) 44–54.

55. T. A. Addona et al., Multi-site assessment of the precision and reproducibility of multiple reaction monitoring-based measurements of proteins in plasma, *Nature Biotechnology* 27 (2009) 633–641; erratum *Nature Biotechnology* 27 (2009) 864.

56. S. S. Rane, A. Ajameri, R. Mody, and P. Padmaja, Development and validation of RP-HPLC and RP-UPLC methods for quantification of erythropoietin formulated with human serum albumin, *Journal of Pharmaceutical Analysis* 2 (2012) 160–165.

57. S. J. Bruce, B. Rochat, A. Beguin, B. Pesse, I. Guessous, O. Boulat, and H. Henry, Analysis and quantification of vitamin D metabolites in serum by ultra-performance liquid chromatography coupled to tandem mass spectrometry and high-resolution mass spectrometry—a method comparison and validation, *Rapid Communications in Mass Spectrometry* 27 (2013) 200–206.

58. R. M. Gathungu, C. C. Flarakos, G. S. Reddy, and P. Vouros, The role of mass spectrometry in the analysis of vitamin D compounds, *Mass Spectrometry Reviews* 32 (2013) 72–86.

59. E. J. Want et al., Ultra performance liquid chromatography-mass spectrometry profiling of bile acid metabolites in biofluids: application to experimental toxicology studies, *Analytical Chemistry* 82 (2010) 5282–5289.

60. B. Shushan, A review of clinical diagnostic applications of liquid chromatography-tandem mass spectrometry, *Mass Spectrometry Reviews* 29 (2010) 930–944.

3 Multilinear Gradient Elution Optimization in Liquid Chromatography

P. Nikitas, A. Pappa-Louisi, and Ch. Zisi

3.1 INTRODUCTION

Gradient elution in liquid chromatography is a powerful technique based on programmed separation modes; therefore, the operation conditions are changed according to a preset program during the elution [1–6]. The main purpose of gradient elution is to accelerate the elution of strongly retained solutes, having the least-retained component well resolved. It may operate in a single mode or multimodes. In single-mode gradient elution, only one separation variable, the mobile phase composition (the most widely used factor affecting retention), the column temperature, the flow rate, or the eluent pH varies with time. The single-program mode is the most common operation of gradient elution since it is easier to control just one variable. In this mode, the performance of gradient elution may be enhanced under multigradient conditions, provided that the relevant theory is available.

In multimode gradient elution, two or more separation variables are used for prediction and optimization. The multimode gradients may be divided into two categories: (a) Gradients related exclusively to the mobile phase composition, that is, the separation variables are the volume fractions ϕ_1, ϕ_2, ... of the constituents (organic modifiers) of the mobile phase or the pH of this phase; and (b) combined gradients of the mobile phase composition with flow rate or column temperature. The fundamental equation for the gradients of the first category is the same with that valid also for single ϕ and pH gradients; it was first proposed by Freiling [7,8] and Drake [9], later by Snyder et al. [2,10–14], whereas more strict derivations of this equation may be found elsewhere [15,16]. However, this fundamental equation is inapplicable for the gradients of the second category, the theory of which was recently developed [17–21].

According to the shape of the gradient program, gradients can be classified as linear (the most common), multilinear, curved (concave or convex), and stepwise. Multilinear gradients (i.e., multiple-linear-segment gradients) have not been widely used in chromatographic applications, in part because of inconvenience in the optimization process. However, the use of multilinear solvent composition ϕ gradients in binary mobile phases [22–43] or in ternary eluents [44], multilinear flow programming technique [45], as well as multilinear combined ϕ/temperature [46] or ϕ/pH gradients in reversed-phase high-performance liquid chromatography (RP-HPLC) [47] can provide significant improvement in resolution of specific samples, although there are limits in the complexity of a gradient program applied to enhance resolution [48]. Some of the gradient profiles used in the applications mentioned were obtained through a trial-and-error optimization approach based on experience and intuition. Significant improvements, however, in the determination of the optimal multilinear gradient profile were obtained through modeling of chromatographic processes and user-friendly optimization routines.

This chapter focuses on multisegment linear gradient optimization strategy in reversed-phase and hydrophilic interaction chromatographic systems since it was recently proved that the same retention models can be applied to both systems [49,50]. For this purpose, the analytical expressions for the retention of solutes obtained under different single- or multimode multilinear gradient conditions are derived based on several models considering the main factors affecting retention, that is, organic modifiers, pH (for ionogenic analytes), temperature, and flow rate. Once these retention models are built with data obtained from sets of carefully designed experiments through the derived analytical expressions embodied to simple fitting algorithms, they can be applied to predict the performance of new gradient conditions as well as to establish the optimal gradient profile by using appropriate optimization algorithms.

3.2 BASIC RELATIONSHIPS

3.2.1 MULTILINEAR GRADIENT PROFILES

A multilinear or segmented gradient profile includes several subsequent linear portions with different slopes. This profile may be completed with isocratic hold periods, most often at the beginning or at the end of elution as well as between linear gradient portions. The general mathematical expression of a multilinear gradient in ϕ, the volume fraction of the organic modifier in the mobile phase, may be expressed as (Figure 3.1)

$$\phi = \begin{cases} \phi_1 = \phi_{in} & t < t_D + t_{in} = t_1 \\ \phi_1 + \lambda_2(t - t_1) & t_1 < t < t_2 \\ \phi_2 + \lambda_3(t - t_2) & t_2 < t < t_3 \\ \cdots & \\ \phi_{n-1} + \lambda_n(t - t_{n-1}) & t_{n-1} < t < t_n \\ \phi_n = \phi_{max} & t > t_n \end{cases} \tag{3.1}$$

where λ_j is the slope of the profile of ϕ versus t at the jth segment, t_D is the dwell time of the chromatographic system (i.e., the time needed for a certain change in the mixer to reach the inlet of the chromatographic column), and t_{in} is the ϕ gradient starting time (i.e., the time the ϕ gradient starts in the pump program). Note that for simplicity we may assume that all slopes are different from zero since a zero slope can be approximated in practice by an infinitesimally small slope.

The mathematical expression of every other separation variable multilinear gradient profile is precisely the same with that of Equation (3.1) with the notable exception

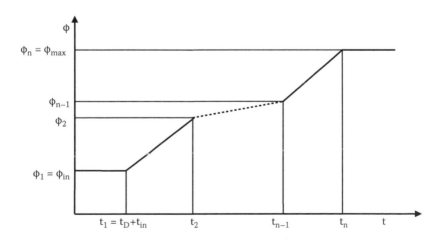

FIGURE 3.1   Schematic multilinear gradient profile of ϕ versus t.

that t_D in Equation (3.1) must be eliminated in the profiles of flow rate, F versus t, and temperature, T versus t, since in these two cases there is no delay between the formation of the gradient and its application to the column.

3.2.2 THE FUNDAMENTAL EQUATION OF GRADIENT ELUTION

The majority of the optimization techniques presuppose the prediction of the elution time t_R of an analyte. This prediction can be based on empirical expressions of
t_R as a function of the characteristics of the gradient profile or on the fundamental
equation of gradient elution. Empirical expressions of t_R are quite effective under
single-gradient profiles [21,51–53], but they have not been used for multilinear gradient profiles until now. The fundamental equation of gradient elution valid under any
type of gradient profile may be expressed as [15,17–21]

$$\int_0^{t_R} \frac{dt_c}{t_0(1+k)} = 1 \tag{3.2}$$

where t_0 is the column holdup time, k is the retention factor, and t_c is a time parameter that has the following physical meaning: Consider a ϕ versus t gradient profile
is formed in the mixer of the chromatographic system. This profile may be approximated by a stepwise curve composed of a great number of infinitesimally small
time steps δt. However, if a step lasts in the mixer δt, inside the column the analyte
feels this step for a time interval δt_c, which is always longer than that of δt. This
is expected because the analyte travels inside the column slower than the mobile
phase. Therefore, a gradient profile ϕ versus t formed in the mixer is no more valid
inside the column where it is transformed to a new profile, ϕ versus t_c. It has been
shown that the time variables t and t_c are interconnected through the following relationship [15]:

$$t_c = \int_0^t \frac{1+k}{k} \, dt \tag{3.3}$$

Two limiting expressions of the fundamental equation of gradient elution are particularly useful:

1. When the mobile phase composition (i.e., the organic modifier content or
   eluent pH) varies at constant flow rate and column temperature, Equation
   (3.2) yields [2,7–16]

$$\int_0^{t_R - t_0} \frac{dt}{t_0 k} = 1 \tag{3.4}$$

This is the fundamental equation for single-gradient elution, and it was
first proposed by Snyder et al. [10–14]

2. If only F or T varies with time, Equation (3.2) is transformed to [2,18,19,45,54]

$$\int_0^{t_R} \frac{dt}{t_0(1+k)} = 1 \tag{3.5}$$

3.2.3 Retention Models

The prediction of the elution time t_R of an analyte through the fundamental equation of gradient elution, Equation (3.2), requires an expression for the retention factor k. In principle, every mathematical expression for k can be used. However, when t_R calculated from the fundamental equation of gradient elution is going to be used in an optimization algorithm, an analytical expression of t_R is usually required. Thus, from the great number of retention models, we have to select those that yield analytical solutions of Equation (3.2) with respect to t_R.

For single-mode gradients in ϕ, the retention models that present the feature discussed are the following: (a) the simple linear model [55]

$$k(\phi) = c_0 + c_1\phi \tag{3.6}$$

which has no practical use; (b) the most commonly used retention model that expresses $\ln k$ linearly on ϕ [2,4,54,56–58]

$$\ln k(\phi) = c_0 - c_1\phi \tag{3.7}$$

and (c) the logarithmic models [24,50]:

$$\ln k = c_0 - c_1 \ln(1 + c_2\phi) \tag{3.8}$$

$$\ln k = c_0 - c_1 \ln \phi \tag{3.9}$$

Note that the last model is a limiting expression of Equation (3.8) when $c_2\phi \gg 1$, and the same is valid for the linear model of Equation (3.7) when $|c_2\phi| \ll 1$. In this case, $\ln(1 + c_2\phi)$ is expanded in a Taylor series, and since $|c_2\phi| \ll 1$, we obtain $\ln(1 + c_2\phi) \approx c_2\phi$, which in combination with Equation (3.8) yields Equation (3.7).

Although Equations (3.7)–(3.9) cover a wide range of applications, the need for more effective models cannot be excluded. Such models are the quadratic model [11,55,59–62],

$$\ln k(\phi) = c_0 + c_1\phi + c_2\phi^2 \tag{3.10}$$

and the rational function model [61–63]:

$$\ln k(\phi) = c_0 - \frac{c_1\phi}{1 + c_2\phi} \tag{3.11}$$

They both exhibit similar applicability, but they cannot be used directly in Equation (3.2) to obtain t_R. In this case, the method developed from our group and described in Section 3.4.2 may be used [28,30]. According to this method, the curve of $\ln k$ versus $\phi(s)$/pH is subdivided into a certain number of linear portions, in each of which Equation (3.7) can be applied.

For gradients in pH, the retention model adopted in the majority of studies of monoprotic analytes is [64–70]

$$k(pH) = \frac{k_0 + k_1 10^{i(pH - pK)}}{1 + 10^{i(pH - pK)}} \tag{3.12}$$

where i is an indicator parameter that is equal to 1 for acids and -1 for bases; k_0, k_1 are the retention factors of the nonionized and the ionized forms of an ionogenic analyte, respectively; and $pK = -\log K$, K being the equilibrium constant of the appropriate acid/base equilibrium. However, Equation (3.12) in combination with Equation (3.4) does not lead to analytical expression for t_R; for this reason, several approximations of Equation (3.12) have been tested [71]. The simplest of them is

$$\frac{1}{k(pH)} = a \left(1 + \frac{b(pH - c)}{\sqrt{1 + (pH - c)^2}} \right) \tag{3.13}$$

For gradients in T, we usually use [72–79]

$$\ln k(T) = c_0 + \frac{c_1}{T} + c_2 g(T) \tag{3.14}$$

where the function $g(T)$ is given by [73]

$$g(T) = \begin{cases} 0 & when & \Delta H^\circ = \Delta H_0 \\ \ln T & when & \Delta H^\circ = \Delta H_0 + \Delta H_1 T \\ T & when & \Delta H^\circ = \Delta H_0 + \Delta H_1 T^2 \\ 1/T^2 & when & \Delta H^\circ = \Delta H_0 + \Delta H_1/T \end{cases} \tag{3.15}$$

Here, ΔH° is the standard enthalpy of the retention process and ΔH_0, ΔH_1 are constants, coefficients of ΔH°.

Many other models have been proposed, and a comprehensive discussion of this issue has been presented [80].

3.3   FITTING AND OPTIMIZATION APPROACHES

The final goal of the separation optimization of a mixture of analytes is to find the optimum gradient profile that yields the best separation of all analytes in the minimum gradient time. The first step toward an optimization technique is the prediction of the elution times of the analytes. This prediction is attained by solving the fundamental equation of gradient elution using a certain retention model. However, all retention models contain adjustable parameters, which should be determined prior to their application for retention prediction. Therefore, a necessary step is the fitting of either the retention model to isocratic data or the expressions of the elution time that arise from the retention model to gradient data.

3.3.1 Fitting Retention Models

As explained, the fitting may be performed using either isocratic or gradient data. In both cases, the objective (cost) function (CF) that should be minimized may be either

$$CF = \sum_{j=1}^{N} (t_{R_j,\exp} - t_{R_j,calc})^2 \tag{3.16}$$

or

$$CF = \sum_{j=1}^{N} (\ln k_{j,\exp} - \ln k_{j,calc})^2 \tag{3.17}$$

where $t_{R_j,\exp}$, $\ln k_{j,\exp}$ are the experimental retention time and the logarithm of the retention factor of a certain solute under the jth measurement, respectively, and $t_{R_j,calc}$, $\ln k_{j,calc}$ are the corresponding calculated values.

It has been shown [21,80] that under isocratic conditions Equations (3.16) and (3.17) do not always give practically converged results. Equation (3.16) reduces the error in high t_R values and increases it in small values of t_R. The behavior of Equation (3.17) is quite the opposite, which seems to have a more balanced behavior, however. In contrast, under gradient-fitting conditions, the two cost functions give similar results. This is clearly shown in Figure 3.2, which depicts

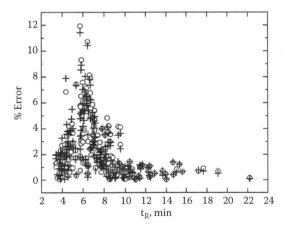

FIGURE 3.2 Percentage absolute error in t_R between experimental and calculated retention times as a function of the experimental t_R of 7 purines, 3 pyrimidines, and 6 nucleosides when the prediction is based on Equation (3.16) (+) and Equation (3.17) (o).

the percentage absolute error in t_R between experimental and calculated retention times as a function of the experimental t_R of seven purines: uric acid, hypoxanthine, xanthine, adenine, theobromine, theophylline, and caffeine; three pyrimidines: cytosine, uracil and thymine; and six nucleosides: cytidine, uridine, inosine, guanosine, thymidine, and adenosine. Data have been taken from Nikitas et al. [24]. We observe that both Equation (3.16) and (3.17) give practically identical results, and that the percentage error is much higher in the range of low t_R values.

The fitting of a model to isocratic data is usually easy and can be performed using commercial software, like the Regression and Solver programs of Excel. Moreover, this kind of fitting gives accurate information about retention in an extended retention range; thus, the determination of the adjustable parameters of a retention model from isocratic data can be used to predict retention reliably in isocratic runs as well as in gradient runs. For these reasons, as well as the fact that the fitting to gradient data may require homemade software, many scientists prefer to use isocratic data to determine the adjustable parameters of a retention model and then use these parameters for prediction and optimization under gradient conditions [29,81–88].

However, despite these advantages, the fitting of a model to gradient data should be preferred since it exhibits (1) less experimental effort; (2) reduction of predictive errors related to nonequilibration of column due to the use of similar experimental conditions [89]; (3) reduction of predictive errors related to the retention model since the distribution of the gradient retention times of a sample solute used for the calculation of the coefficients of a retention model is much narrower than that if isocratic data are used in the fitting procedure. However, a basic prerequisite to obtain high-quality gradient predictions based on gradient data is the following: The predicted retention time of each solute should lie within the corresponding values of the gradient data used to determine the retention model. If this prerequisite is not valid, in the fitting procedure we should add more initial gradient data or replace some gradient data by proper isocratic ones [30].

Finally, much attention is usually paid on statistical tests for the significance of the various adjustable parameters of a retention model or for the choice of the proper retention model [28,90,91]. However, in our opinion, this is a rather overestimated issue unless it is related to overfitting problems. That is, if there are no overfitting problems, there is no need to reduce the number of the adjustable parameters and keeping only the statistical significant ones because the ultimate criterion for a retention model is that the percentage error between calculated and experimental retention times should be as small as possible.

3.3.1.1 Fitting Algorithms for Gradient Data

The cost functions, Equations (3.16) and (3.17), are differentiable functions; therefore, in principle, the fitting problem can be solved using algorithms based on the Levenberg–Marquardt (LM) method [92–94]. However, the direct application of the LM algorithm exhibits several problems, the most important of which is the trapping of the algorithm into local minima other than the global one. For this reason, variations of this algorithm have been proposed [28,95].

A modification of the LM algorithm (RND_LM) arises if it is combined with an initial random (RND) search of the domain adopted for the adjustable parameters of the retention model [28,95]. Thus, a number of vectors of adjustable parameters are randomly selected from the search domain, and the vector that yields the minimum value of the objective function is selected; its coordinates are used as initial estimates in the LM algorithm, which determines the minimum of the objective function. In another modification (R_LM), the LM algorithm starts with an initial vector of adjustable parameters randomly selected from the search domain, and the LM method determines the local minimum, which is stored. Then, a new initial vector of adjustable parameters is randomly selected, and the whole algorithm is repeated (R) for a preset number of iterations. The minimum of the stored local minima is determined, and it presumably corresponds to the global minimum of the objective function.

Based on the RND_LM algorithm, several macros have been written in VBA (Visual Basic of Applications) that can run in Microsoft Excel [95–98]. In addition, for the easy implementation of the RND_LM and R_LM algorithms, homemade software was written in C++ to run under Windows. The exe files of these algorithms are free and available on request from the authors.

Apart from the LM-based algorithms, genetic algorithms (GAs) [99,100], algorithms based on the descent (D) method [94,101] as well as combinations of GA and D with LM have been also proposed to solve fitting problems [102,103]. However, the LM-based algorithms are still a good tool for solving fitting problems since they are computationally simple and fast. In contrast, GA and D algorithms do not offer any advantage, whereas they are on average much slower than the algorithms based on LM.

3.3.2 OPTIMIZATION TECHNIQUES

The optimization techniques adopted to find the optimum gradient profile that yields the best separation of all analytes of a complex mixture to the minimum gradient time may be based on single- or multiobjective optimization criteria. The mathematical expression of the objective functions are based on the resolution $R_{S,ij}$ of adjacent solutes i and j defined from

$$R_{S,ij} = 2\frac{|t_{R,j} - t_{R,i}|}{w_{bj} + w_{bi}} \tag{3.18}$$

where w_{bi}, w_{bj} are the peak widths at base. Alternatively, in place of $R_{S,ij}$ we can use retention time absolute differences:

$$\delta t_{R,ij} = |t_{R,i} - t_{R,j}| \tag{3.19}$$

This choice is suited when the peaks are relatively narrow with regard to the peak distance.

In the single-objective optimization problem, we use just one objective (cost) function to minimize or maximize, depending on the formulation of this function. Two typically adopted single-objective functions are

$$\text{Minimize } CF = t_{R,\max} \text{ subject to } R_{s,\min} > R_{\min} \tag{3.20}$$

$$\text{Maximize } CF = \delta t_{\min} \text{ subject to } t_{R,\max} < t_{g,\max} \tag{3.21}$$

where $t_{R,\max}$ is the elution time of the most distant solute, $t_{g,\max}$ is the maximum gradient elution time preset by the researcher, $R_{s,\min}$ is the smallest $R_{S,ij}$ value, $R_{\min}$ is a preset constant usually equal to 1.5, and $\delta t_{\min}$ is the minimum value of $\delta t_{R,ij}$ or $R_{S,ij}$.

For multiobjective optimization, we may use the weighted sum method, in which all the objective functions are summed provided that each one is multiplied by a proper weighting factor. The first objective function of this form was suggested by Berridge [104]:

$$\text{Minimize } CF = -\sum_i R_{S,ij} - P^{w_1} + w_2 \mid t_{g,\max} - t_{R,\max} \mid + w_3 \mid t_{R,0} - t_{R,1} \mid \tag{3.22}$$

where P is the number of peaks detected, $t_{R,0}$ is the minimum retention time desired for the first eluted peak, and $t_{R,1}$ is the real retention time for that first peak. Factors w_1, w_2, w_3 are operator-selectable weightings that, according to Berridge, are usually set to values between 0 and 3. A simpler expression of this type is

$$\text{Minimize } CF = -\sum_{i=1}^{m} w_i \delta t_i + \mid t_{R,\max} - t_{g,\max} \mid \tag{3.23}$$

where m is a small integer, say 3, and in this case δt_1, δt_2, δt_3 are the first three minimum values of $\delta t_{R,ij}$ or $R_{S,ij}$.

An alternative approach for multiobjective optimization is to examine more than one objective function separately:

$$\text{Maximize/minimize } CF_m = f_m(x), \, m = 1, 2, ..., M \tag{3.24}$$

subject to certain constraints. In this case, methods of multicriteria decision making should be applied for estimation of the optimum. The most popular of them is Pareto optimality [105–109], in which one compares all the experimental results with each other. A point is called Pareto optimal if there is no other solution that has a better result in one objective without having a worse result in another objective. In fact, there is not one optimum point, but there are several Pareto-optimal points. Among these Pareto-optimal solutions, one or several solutions are selected for the optimum separation of the sample mixture by secondary or personal criteria.

Other approaches, like the use of the desirability function or the multicriteria decision-making method proposed by Derringer, have also been used for gradient elution optimization in reversed-phase liquid chromatography [110–113].

3.3.2.1 Optimization Algorithms

All the objective functions used for optimization are not differentiable functions; therefore, the LM-based algorithms adopted in the fitting problem cannot be used for optimization. For this reason, GAs [30,114,115] as well as algorithms based on the descent method [28,44,53] have been proposed and used for the determination of the optimum gradient.

On this issue, we have found [30] that the classical GA suggested by Michalewicz [99] with linear scaling [100], combined crossover, and Gaussian mutation [99] performs very well when using population size 100, crossover probability 0.8, and probability of mutation 0.02. In what concerns the algorithms based on the descent method, their performance is considerably increased if they search thoroughly the area close to a current solution but at the same time explore far distant areas to determine the possibility of better solutions, which might lead to the global minimum.

However, when the number of the separation variables is relatively small, more simple techniques, like a grid [17,20,116] or a totally random search [24], are quite effective. In this case, we may adopt the objective function of Equation (3.23) with $w_1 = 1$ and $w_2 = w_3 = 0$ to construct the Pareto front. In fact, a simple plot of δt_1 and $t_{R,max}$ is enough to determine the optimum solution(s). If no practical solution can be found, one may examine the plots δt_2 versus $t_{R,max}$ or δt_3 versus $t_{R,max}$. Such a simple plot of $-\delta t_1$ versus $t_{R,max}$ used for optimizing the gradient profile in Nikitas et al. [24] is shown in Figure 3.3.

3.4   RETENTION PREDICTION AND SEPARATION OPTIMIZATION UNDER MULTILINEAR GRADIENT ELUTION CONDITIONS

In this section, different approaches for both prediction and optimization under multilinear gradient elution conditions are presented in detail. Specifically, expressions for the retention time of solutes are derived when (a) the content of one or more organic modifiers in the mobile phase, (b) the eluent pH, (c) both content of organic

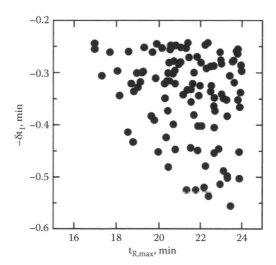

FIGURE 3.3   Plot of $-\delta t_1$ versus $t_{R,max}$ used for optimizing the gradient profile in Nikitas et al. [24].

modifier and eluent pH, (d) the column temperature T, and (e) the flow rate F vary with time according to a multilinear profile during a gradient run.

3.4.1 ORGANIC MODIFIER GRADIENTS

When a multilinear gradient in ϕ is applied, an analyte may be eluted in the first or the last isocratic portions or in the jth linear segment. For the expression of t_R, we first examine the retention model of Equation (3.8) since it generates both the simple linear model of Equation (3.7) and the simple logarithmic model of Equation (3.9). In the next step, we are presenting the expressions of t_R based on Equation (3.7) and Equation (3.9) since the treatment is almost identical.

The analyte is eluted in the first isocratic portion when

$$I_1 = \int_0^{t_{in}+t_D} \frac{dt}{t_0 k} = \frac{t_{in}+t_D}{t_0 k_{in}} = \int_0^{t_R-t_0} \frac{dt}{t_0 k} + \int_{t_R-t_0}^{t_{in}+t_D} \frac{dt}{t_0 k} \geq 1 \qquad (3.25)$$

since, according to Equation (3.4), the integral from 0 to t_R - t_0 is always equal to 1, and the integral from t_R - t_0 to t_D + t_1 is a positive quantity. When the condition $I_1 \geq 1$ is valid, the retention time is calculated from the well-known expression of isocratic elution:

$$t_R = t_0(1 + k_{in}) \qquad (3.26)$$

which results directly from Equation (3.4) by integration under constant $k = k(\phi_{in}) = k_{in}$.

When the analyte is eluted in the jth linear segment ($j > 1$), that is, between t_{j-1} and t_j, the condition is

$$\int_0^{t_{j-1}} \frac{dt}{t_0 k} < 1 \text{ and } \int_0^{t_j} \frac{dt}{t_0 k} \geq 1 \qquad (3.27)$$

The last integral may be expressed as

$$\int_0^{t_j} \frac{dt}{t_0 k} = \int_0^{t_1} \frac{dt}{t_0 k} + \int_{t_1}^{t_2} \frac{dt}{t_0 k} + \cdots + \int_{t_{j-1}}^{t_j} \frac{dt}{t_0 k} = I_1 + I_2 + \cdots + I_j \qquad (3.28)$$

Therefore, the condition for the analyte to be eluted in the jth linear segment may be expressed as

$$I_1 + I_2 + \cdots + I_{j-1} < 1 \text{ and } I_1 + I_2 + \cdots + I_{j-1} + I_j \geq 1 \qquad (3.29)$$

where I_1 is calculated from Equation (3.25) and I_j, $j > 1$, from

$$I_j = \int_{t_{j-1}}^{t_j} \frac{dt}{t_0 k} = \int_{t_{j-1}}^{t_j} \frac{(1+c_2\phi)^{c_1} dt}{t_0 e^{c_0}} = \frac{(1+c_2\phi_j)^{c_1+1} - (1+c_2\phi_{j-1})^{c_1+1}}{(c_1+1)c_2 t_0 \lambda_j e^{c_0}} \qquad (3.30)$$

since

$$\phi = \phi_{j-1} + \lambda_j (t - t_{j-1}).$$

In this case, the retention time t_R of the analyte is calculated from

$$\int_{t_{j-1}}^{t_R - t_0} \frac{dt}{t_0 k} = 1 - \int_0^{t_{j-1}} \frac{dt}{t_0 k} = 1 - I_1 - I_2 - \cdots - I_{j-1} \tag{3.31}$$

which results in

$$t_R = t_0 + t_{j-1} + \frac{B_j^{\frac{1}{c_1+1}} - 1 - c_2 \phi_{j-1}}{c_2 \lambda_j} \tag{3.32}$$

where

$$B_j = (c_1 + 1) c_2 t_0 e^{c_0} \lambda_j (1 - I_1 - I_2 - \cdots - I_{j-1}) + (1 + c_2 \phi_{j-1})^{c_1+1} \tag{3.33}$$

Finally, when the analyte is eluted in the last isocratic portion, that is, when $\phi = \phi_n = \phi_{max}$, the condition is

$$\int_0^{t_n} \frac{dt}{t_0 k} < 1 \implies I_1 + I_2 + \cdots + I_{n-1} + I_n < 1 \tag{3.34}$$

and the retention time t_R arises from

$$\int_{t_n}^{t_R - t_0} \frac{dt}{t_0 k} = 1 - I_1 - I_2 - \cdots - I_n \tag{3.35}$$

which yields

$$t_R = t_0 + t_n + t_0 k_{max} (1 - I_1 - I_2 - \cdots - I_n) \tag{3.36}$$

since $k = k(\phi_{max}) = k_{max}$ is constant. Table 3.1 summarizes these results.

If we work as discussed using the retention models of Equations (3.9) and (3.7), we obtain that Equations (3.26) and (3.36) are still valid when the analyte is eluted in the first and the last isocratic portions, respectively. However, I_2, I_3, ... are no more defined from Equation (3.30). In particular, the retention model of Equation (3.9) yields for t_R the expression presented in Table 3.1, where the symbols I_j and B_j are defined from

$$I_j = \frac{\phi_j^{c_1+1} - \phi_{j-1}^{c_1+1}}{(c_1 + 1) t_0 e^{c_0} \lambda_j}, \, j > 1 \tag{3.37}$$

TABLE 3.1

Expressions of the Retention Time under Multilinear Organic Modifier Gradients When the Retention Model Is Given by Equation (3.7), (3.8), or (3.9)

Section	Condition	t_R
1	$I_1 \geq 1$	$t_R = t_0(1 + k_{in})$
j	$I_1 + I_2 + \cdots + I_{j-1} < 1$	I. Equation (3.7):
	$I_1 + I_2 + \cdots + I_{j-1} + I_j \geq 1$	$t_R = t_0 + \dfrac{\ln\{(1 - I_1 - \cdots - I_{j-1})/J_j + e^{c_1 \lambda_j t_{j-1}}\}}{c_1 \lambda_j}$
		II. Equation (3.8):
		$t_R = t_0 + t_{j-1} + \dfrac{B_j^{\frac{1}{c_1+1}} - 1 - c_2 \phi_{j-1}}{c_2 \lambda_j}$
		III. Equation (3.9):
		$t_R = t_0 + t_{j-1} + \dfrac{B_j^{\frac{1}{c_1+1}} - \phi_{j-1}}{\lambda_j}$
> n	$I_1 + I_2 + \cdots + I_{n-1} + I_n < 1$	$t_R = t_0 + t_n + t_0 k_{\max}(1 - I_1 - I_2 - \cdots - I_n)$

and

$$B_j = (c_1 + 1)t_0 e^{c_0} \lambda_j (1 - I_1 - \cdots - I_{j-1}) + \phi_{j-1}^{c_1+1} \tag{3.38}$$

In what concerns the simple linear model of Equation (3.7), it yields for t_R the expression shown also in Table 3.1, where

$$I_j = \frac{e^{c_1 \phi_j} - e^{c_1 \phi_{j-1}}}{t_0 e^{c_0} c_1 \lambda_j}, j > 1 \text{ and } J_j = \frac{e^{c_1 \phi_{j-1} - c_1 \lambda_j t_{j-1}}}{t_0 e^{c_0} c_1 \lambda_j} \tag{3.39}$$

The theory presented in this section was developed in Nikitas et al. [24], and it was tested using a mixture of 16 model compounds chosen among purines, pyrimidine, and nucleosides in eluting systems modified by acetonitrile. The retention of solutes was investigated under 16 different gradient elution programs: 5 simple monolinear gradient profiles without any initial isocratic part, 7 simple linear gradient profiles with an initial isocratic segment, and 4 multilinear programs encompassing an initial isocratic part followed by a bilinear variation of organic modifier content. The obtained experimental retention data were used for a detailed examination of the fitting and prediction performance of Equations (3.7), (3.8), and (3.9) through their corresponding expressions of t_R presented in Table 3.1.

A comparison of the validity of the three models in the description of the solute retention at each gradient run is shown in Figure 3.4. From this figure, it is clear that

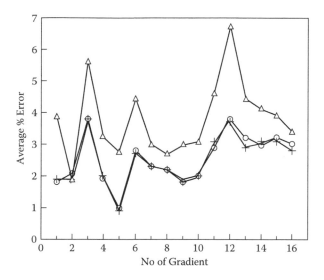

FIGURE 3.4 Average absolute percentage errors between experimental and calculated retention data obtained from Equation (3.7) (Δ), Equation (3.8) (+), and Equation (3.9) (o) at each gradient run. (Adapted and modified from P. Nikitas, A. Pappa-Louisi, P. Agrafiotou, and A. Mansour, *J. Chromatogr. A,* 1218: 5658, 2011, with permission.)

both Equation (3.8) and (3.9) exhibit a quite satisfactory and practically identical performance concerning the gradient data retention prediction, whereas the average percentage errors between experimental and calculated retention data by means of Equation (3.7) at each gradient run are systematically greater than those obtained from the logarithmic models. Consequently, the simple two-parameter logarithmic Equation (3.7) seems to be the proper choice for the retention modeling of the solutes under study, and for this reason, this model was used in the optimization algorithm to determine the linear or multilinear gradient profile that is expected to lead to optimum separation of the sample of interest. The optimum gradient profile determined for a total elution time equal to 17 min is depicted in Figure 3.5 along with the chromatogram recorded under these optimal gradient conditions. Indeed, the chromatogram shows a perfect resolution of the mixture of analytes used. Note that the same optimum was obtained when we used the retention model of Equation (3.8) instead of Equation (3.9) in the optimization algorithm. Consequently, there is no reason to use the three-parameter logarithmic expression for the retention behavior at least of the analytes adopted in Nikitas et al. [24].

The analytical solutions of the fundamental equation appeared in Table 3.1 for Equation (3.7) were also applied to hydrophilic interaction chromatography (HILIC) gradient elution [49]. To investigate the applicability of the previous analytical expressions to a HILIC system, the retention of 21 amino acids was studied with an ACQUITY Ultra Performance Liquid Chromatographic System (Waters Corporation, Milford, MA, USA) under 14 different gradient elution programs over 30 min. The concentration of mobile phases was changed by increasing eluent B (which consisted of 2% acetonitrile, 49% methanol, and 49% water) over eluent A

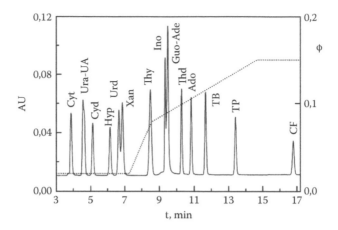

FIGURE 3.5 UV detected chromatogram of a 16-component mixture of solutes: Cytosine (Cyt), uracil (Ura), uric acid (UA), cytidine (Cyd), hypoxanthine (Hyp), uridine (Urd), xanthine (Xan), thymine (Thy), inosine (Ino), guanosine (Guo), adenine (Ade), thymidine (Thd), adenosine (Ado), theobromine (TB), theophylline (TP), and caffeine (CF). The chromatogram was obtained using the optimal gradient profile. The dotted line shows the optimal variation of organic content of mobile phase, ϕ versus t, when it reaches the UV detector. (Reprinted from P. Nikitas, A. Pappa-Louisi, P. Agrafiotou, and A. Mansour, *J. Chromatogr. A*, 1218: 5658, 2011, with permission.)

(which consisted of 96% acetonitrile, 2% methanol, 2% water). In all gradient programs applied, the final proportion of eluent B in the mobile phase was 60% (i.e., $\phi_B = 0.6$) and included a series of nine gradient bilinear programs encompassing two linear portions of different slopes of ϕ_B plus five curve profiles that are available by the Waters Acquity UPLC MassLynx software (two concave, one simple linear, and two convex). The gradient profiles applied are shown in Figure 3.6.

If the analytical expressions of Table 3.1 concerning Equation (3.7) are applied to HILIC systems, ϕ may represent the volume fraction of the stronger modifier in the eluent system (i.e., of the water). However, the mobile phases used [49] are ternary mixtures that consisted of acetonitrile, methanol, and water, in which the ratio of volume fractions of "protic modifiers" (i.e., of water and methanol) was kept constant (1:1). Consequently, when Equation (3.7) is applied to such ternary mobile phases in HILIC separations, ϕ may denote the total water-methanol content in the eluent, considering the mixture of water-methanol (1:1 v/v) as the strong eluting solvent. For more convenience, instead of $\phi_{water-methanol}$, the programmed volume fraction contribution of line B, ϕ_B, in the mobile phase can be used during each gradient run since this line contains the stronger eluent, which consisted of 2% acetonitrile and 98% methanol-water (1:1 v/v), in comparison to that in line A, which consisted of 96% acetonitrile and 4% methanol-water (1:1 v/v). Moreover, to apply the expressions that appear in Table 3.1 to gradients with a curved shape, the gradient profiles should be approximated by multilinear shapes such as the six-linear profiles depicted in Figure 3.6. Note that starting from data from at least two gradient runs, the prediction of solute retention obtained under all the rest of the gradients was

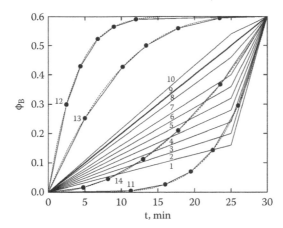

FIGURE 3.6 Fourteen different gradient elution profiles applied on the Hypersil GOLD AX column under HILIC conditions. 1–8 and 10, bilinear; 9, monolinear; 11 and 14, concave; 12 and 13, convex. (Reprinted from H. Gika, G. Theodoridis, F. Mattivi, U. Vrhovsek, and A. Pappa-Louisi, *J. Sep. Sci.*, 35: 376–383, 2012, with permission.)

excellent (the overall predicted error was only 1.2%), even when curved gradient profiles approximated by multilinear ones were used. Consequently, it was proved [49] that the simple linear retention model, Equation (3.7), is a proper choice for gradient retention predictions in HILIC separations through the analytical solution of the fundamental equation of the multilinear gradient elution derived for reversed-phase systems. It should be pointed out here that we did not proceed [49] with the determination of the optimal multilinear gradient profile by using an appropriate optimization algorithm since this was not feasible for the experimental system under investigation. However, the potential for using the above presented retention modeling for HILIC separation optimizations in the future remains.

Finally, in a recent work [50], the retention time predictive ability of the linear retention model, Equation (3.7), as well as of the simple logarithmic retention model of Equation (3.9), was further explored in a HILIC system using a different type of stationary phase, different mobile phase system (which included [A] acetonitrile-water (95:5, v:v), 0.1% formic acid, 0.075% ammonium hydroxide as a weak eluting solvent and [B] acetonitrile- water (2:98, v:v), 0.2% formic acid, 0.1% ammonium hydroxide as a strong eluting solvent) and a set of analytes consisting of 23 polar metabolites from various classes of compounds. The experimental procedure that involved retention data acquisition was based on two different types of gradient programs. In the first type, the gradient profiles were taken from two linear gradient segments without any initial isocratic part (Figure 3.7a). In the second gradient type, there was an initial isocratic part of 1 to 4 min followed by two linear gradient variations between the same values of ϕ_B (0 → 0.28 and 0.28 → 0.6, respectively) of different durations (Figure 3.7b). In total, the series of analyses was performed under ten gradient programs of the first type and under seven of the second type. Starting from at least three gradient

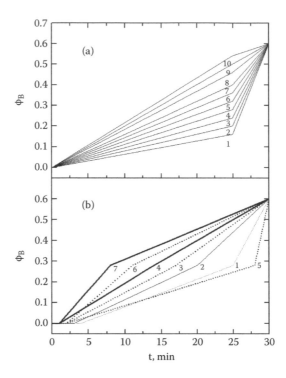

FIGURE 3.7   Applied gradient elution profiles on the ACQUITY UPLC BEH amide column under HILIC conditions: (a) 10 gradient profiles without initial isocratic part; (b) 7 gradient profiles with an initial isocratic part of 1 to 4 min. (Reprinted from H. Gika, G. Theodoridis, F. Mattivi, U. Vrhovsek, and A. Pappa-Louisi, *Anal. Bioanal. Chem.*, 404: 701, 2012, with permission.)

runs, the prediction of analyte retention was satisfactory for all gradient programs tested, providing useful evidence of the value of such retention time prediction methodologies in HILIC systems. A detailed comparison of the predictive ability of both retention models, Equations (3.7) and (3.9), is given in Figure 3.8.

3.4.2   Nikitas–Pappa's Approach for Organic Modifiers and pH Gradients

The treatments discussed are based on the selection of the proper retention model that yields analytical expressions for the retention time t_R of an analyte. This restriction excludes use of many retention models, possibly more effective than those of Equations (3.7)–(3.9). To overcome this problem, we have developed a new approach [28,30,116], a generalization of which is presented here.

Every nonlinear curve of ln k versus t can always be subdivided into m linear portions, allowing for an analytical expression of t_R. In more detail, consider a gradient profile consisting of the multilinear curves $X_1 = g_1(t)$, $X_2 = g_2(t)$, ... , $X_q = g_q(t)$, where $X_1, X_2, ... , X_q$ may be volume fractions of the organic modifiers or the pH of the mobile phase. These multilinear curves may be expressed as vectors with common

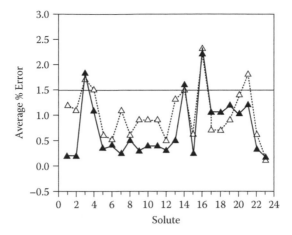

FIGURE 3.8   Mean absolute percentage errors between experimental and calculated retention data of each solute obtained from Equation (3.7) (Δ - dotted line) and from Equation (3.9) ($\blacktriangle$ - solid line) at all seventeen gradient runs tested. (Reprinted from H. Gika, G. Theodoridis, F. Mattivi, U. Vrhovsek, and A. Pappa-Louisi, *Anal. Bioanal. Chem.*, 404: 701, 2012, with permission.)

time components $(X_{0i}, X_{1i}, \ldots, X_{pi}, T_1, T_2, \ldots, T_p)$, where $i = 1, 2, \ldots, n$ and $p \geq q$, like the vectors $(X_{01}, X_{11}, X_{21}, X_{31}, X_{41}, T_1, T_2, T_3, T_4, T_5)$ and $(X_{02}, X_{12}, X_{22}, X_{32}, X_{42}, T_1, T_2, T_3, T_4, T_5)$ in Figure 3.9.

Now, each of these portions (T_i, T_{i+1}) is divided to a certain number of subportions, where we assume that $\ln k$ varies linearly with t. This division results in a further division of the time axis, $t_1, t_2, \ldots, t_n$ (see example in Figure 3.9). At each of these sections, say in the range $[t_{j-1}, t_j]$, $\ln k_j$ varies linearly with t:

$$\ln k_j = \ln A_j - B_j t \quad \text{when } t_{j-1} \leq t \leq t_j \tag{3.40}$$

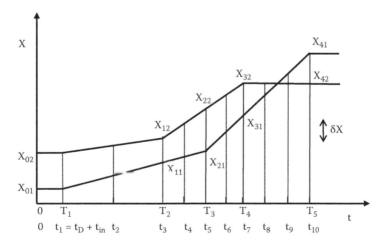

FIGURE 3.9   Schematic multilinear gradient profiles of two separation variables X_1 and X_2 versus t.

Note that

$$\ln k(X_1, X_2, \dots, X_n) = \ln k(t) = f(t) \tag{3.41}$$

and if function $f(t)$ is known from the isocratic or gradient properties of the sample solute, constants A_j and B_j may be calculated from

$$B_j = -\frac{f(t_j) - f(t_{j-1})}{t_j - t_{j-1}} \tag{3.42}$$

and

$$\ln A_j = f(t_{j-1}) + B_j t_{j-1} \Rightarrow A_j = e^{f(t_{j-1}) + B_j t_{j-1}} \tag{3.43}$$

The analyte is eluted in the jth section, that is, in the range $[t_{j-1}, t_j]$, when conditions expressed by Equation (3.29) are valid, where now

$$I_j = \frac{1}{t_0} \int_{t_{j-1}}^{t_j} \frac{e^{B_j t} dt}{A_j} = \frac{e^{B_j t_j} - e^{B_j t_{j-1}}}{t_0 A_j B_j} \tag{3.44}$$

Equation (3.44) is valid for $j > 1$, whereas I_1 is still given by Equation (3.25). In this case, the retention time is calculated from

$$\int_{t_{j-1}}^{t_R - t_0} \frac{e^{B_j t} dt}{A_j} = (1 - I_0 - \dots - I_{j-1}) t_0 \tag{3.45}$$

which yields

$$t_R = t_0 + \frac{D_j}{B_j} \tag{3.46}$$

where

$$D_j = \ln \left(A_j B_j (1 - I_1 - \dots - I_{j-1}) t_0 + e^{B_j t_{j-1}} \right) \tag{3.47}$$

Equation (3.46) is valid when B_j is different from zero. If $B_j = 0$, then Equation (3.45) results in

$$t_R = t_0 + t_{j-1} + A_j (1 - I_1 - \dots - I_{j-1}) t_0 \tag{3.48}$$

Finally, when the analyte is eluted in the first or last isocratic portions of the gradient profile, the conditions and the expressions for t_R found in the previous section are still valid. These results are summarized in Table 3.2.

TABLE 3.2

Expressions of the Retention Time under Multilinear Gradient Conditions According to Nikitas–Pappa's Approach

Section	Condition	t_R
1	$I_1 \geq 1$	$t_R = t_0(1 + k_{in})$
j	$I_1 + I_2 + \cdots + I_{j-1} < 1$	I. $\ln k_j$ varies linearly with time:
	$I_1 + I_2 + \cdots + I_{j-1} + I_j \geq 1$	$B_j \neq 0$: $t_R = t_0 + \dfrac{D_j}{B_j}$
		$B_j = 0$: $t_R = t_0 + t_{j-1} + A_j(1 - I_1 - \ldots - I_{j-1})t_0$
		II. k_j varies linearly with time:
		$B_j \neq 0$: $t_R = t_0 + \dfrac{e^{D_j} - A_j}{B_j}$
		$B_j = 0$: $t_R = t_0 + t_{j-1} + A_j(1 - I_1 - \ldots - I_{j-1})t_0$
> n	$I_1 + I_2 + \cdots + I_{n-1} + I_n < 1$	$t_R = t_0 + t_n + t_0 k_{max}(1 - I_1 - I_2 - \cdots - I_n)$

If we adopt that k_j and not $\ln k_j$ varies linearly with t, possibly when we have pH gradients, then

$$k_j = \ln A_j - B_j t \quad \text{when } t_{j-1} \leq t \leq t_j \tag{3.49}$$

and Equations (3.42) and (3.43) are replaced by

$$B_j = \frac{f(t_j) - f(t_{j-1})}{t_j - t_{j-1}} \quad \text{and} \quad A_j = f(t_{j-1}) - B_j t_{j-1} \tag{3.50}$$

In this case, I_j is defined from

$$I_j = \frac{1}{t_0} \int_{t_{j-1}}^{t_j} \frac{dt}{A_j + B_j t} = \frac{1}{t_0 B_j} \ln \frac{A_j + B_j t_j}{A_j + B_j t_{j-1}} \tag{3.51}$$

and the obtained retention times are given in Table 3.2, where D_j is defined from

$$D_j = \ln(A_j + B_j t_{j-1}) + B_j(1 - I_1 - \ldots - I_{j-1})t_0 \tag{3.52}$$

Two applications of multilinear ϕ gradient elution optimization based on the approach presented have been carried out [28,30]. In these applications, the rational function model expressed by Equation (3.11) was adopted for the retention behavior of solutes under investigation, which were a mixture of 13 o-phthalaldehyde (OPA) derivatives of amino acids [28] or 15 underivatized amino acids [30] with mobile

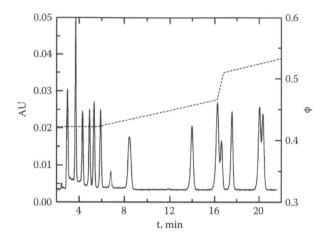

FIGURE 3.10 UV-detected chromatogram of OPA derivatives of amino acids Arg, Gln, Ser, Glu, Thr, Dopa, Ala, Met, Val, Trp, Phe, Ile, and Leu (from left to right) under the optimal gradient conditions. The small peak shown by dots corresponds to OPA, and the crooked line shows the optimum variation pattern of ϕ versus t when it reaches the UV detector. (Reprinted from P. Nikitas, A. Pappa-Louisi, and A. Papageorgiou, *J. Chromatogr. A* 1157: 178, 2007, with permission.)

phases modified by acetonitrile. In both publications, the adjustable parameters of Equation (3.11) were determined by fitting to gradient data through the expressions of t_R that appear in Table 3.2 when ln k_j varies linearly with time. The obtained parameters were further used for the prediction of the best gradient profile that yields the best separation of analytes by means of an algorithm based on the descent method [28] or a GA [30]. The chromatogram of Figure 3.10 shows the separation of 13 OPA derivatives of amino acids achieved by the determined optimum multilinear gradient profile for a total elution time equal to 20 min. Note that there was a very good prediction of the solute retention under optimum conditions shown in Figure 3.10 since the average percentage error of the predicted values was only 2.7% [28]. A similarly good retention prediction under optimum multilinear gradient conditions was also achieved for the separation of 15 underivatized amino acids shown in Figure 3.11.

Moreover, the multilinear gradient approach presented has been applied in ternary mobile phases [44]. For the evaluation of the theory and the performance of the various fitting and optimization algorithms in this case, we used 13 OPA derivatives of amino acids with mobile phases modified by acetonitrile and methanol. It was shown that the theory can lead to high-quality predictions of the retention times and of the optimal multilinear gradient conditions provided that we used the most commonly adopted expression for ln k in ternary solvent systems [117,118],

$$\ln k = a + a_1\phi_a + a_2\phi_b + a_3\phi_a^2 + a_4\phi_b^2 + a_5\phi_a\phi_b \qquad (3.53)$$

and the adjustable parameters of this expression are determined from ternary isocratic data. In Equation (3.53), ϕ_a and ϕ_b are the volume fractions of the two organic

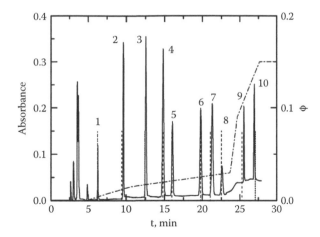

FIGURE 3.11 UV-detected chromatogram of (1) met, (2) dopa, (3) tyr, (4) m-tyr, (5) me-dopa, (6) phe, (7) me-tyr, (8) 5htp, (9) n-tyr, and (10) trp under optimal gradient conditions. Vertical lines indicate the predicted retention times, whereas the dash-dotted line shows the optimal multilinear gradient profile when it reaches the UV detector. The order of the amino acids eluted in the first 6 min of the chromatogram is his, car, cre, crn, and hcy. (Reprinted from P. Nikitas, A. Pappa-Louisi, and P. Agrafiotou, *J. Chromatogr. A*, 1120: 299, 2006, with permission.)

modifiers in the mobile phase. The optimum gradient obtained for a total elution time equal to 20 min is shown in Figure 3.12, where the acceptable resolution achieved for the solute mixture is also clear.

3.4.3 ORGANIC MODIFIER GRADIENTS IN DIFFERENT ELUENT pHs

In a recent article [119], an interesting study of ionizable analytes under RP-HPLC organic modifier gradient elution was presented. In particular, 12 acid–base compounds with pK_a values ranging from 4 to 9 were examined in three different buffered mobile phases (pH 5, pH 7, and pH 9) under ϕ gradient conditions with acetonitrile as the organic modifier. In this work, an equation describing the retention of ionizable analytes was combined with three different models of one, two, or three fitting parameters. All models were tested under 16 different gradient patterns—4 linear gradients, 4 concave gradients, 4 convex gradients, and 4 combinations among them—from which one was a multilinear gradient, and the agreement between the experimental and calculated retention times was found to be almost good for all models, especially for the three-parameter model.

In this work, the authors solved the fundamental equation of gradient elution using a stepwise method proposed in Nikitas and Pappa-Louisi [15]. According to this method, the time axis is divided into a great number of time steps, each one with 0.01- or 0.02-min time duration. The method is quite effective for retention prediction, but it becomes almost inapplicable for optimization procedures under gradient conditions due to a significant increase in the computational time. To overcome this

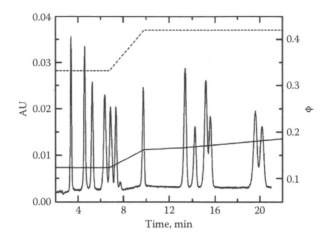

FIGURE 3.12 UV-detected chromatogram of OPA derivatives of amino acids Arg, Gln, Ser, Glu, Thr, Dopa, Ala, Met, Trp, Val, Phe, Ile, and Leu (from left to right) under the optimal ternary gradient determined by programmable optimization. The crooked lines (- - -) and (—) show the variation pattern of $\phi_{Acetonitrile}$ versus t and $\phi_{Methanol}$ versus t, respectively, when they reach the UV detector. (Reprinted from A. Pappa-Louisi, P. Nikitas, and A. Papageorgiou, *J. Chromatogr. A*, 1166: 126, 2007, with permission.)

problem, we may modify the original treatment that suits with the approach presented in Section 3.4.1.

Thus, when ionogenic monoprotic analytes are eluted during multilinear ϕ gradient runs performed in different eluent pHs, the profile of the gradients is still expressed by Equation (3.1), whereas the retention model is described by Equation (3.12), which can be written as

$$k = k_0 \frac{1 + f10^{i(pH-pK)}}{1 + 10^{i(pH-pK)}} \tag{3.54}$$

where

$$f = k_1 / k_0 \tag{3.55}$$

In this approach, following the approximation suggested [119], we assume that the ratio f is constant during the ϕ gradient, whereas k_0 may be expressed by one of the models of Equations (3.7), (3.8), or (3.9). Thus, Equation (3.54) yields

$$\ln k = \ln k_0 + \ln D \tag{3.56}$$

where

$$D = \frac{1 + f10^{i(pH-pK)}}{1 + 10^{i(pH-pK)}} \tag{3.57}$$

TABLE 3.3

Expressions of the Retention Time under Multilinear Organic Modifier ϕ Gradients in Different Eluent pHs When the Retention Model Is Given by Equation (3.54), Whereas $\ln k_0$ Is Given by One of Equations (3.7), (3.8), and (3.9)

Section	Condition	t_R
1	$I_1 \geq 1$	$t_R = t_0(1 + k_{in})$
j	$I_1 + I_2 + \cdots + I_{j-1} < 1$ $I_1 + I_2 + \cdots + I_{j-1} + I_j \geq 1$	I. Equation (3.7): $t_R = t_0 + \dfrac{\ln A - a_1}{a_2}$ II. Equations (3.8), (3.9): $t_R = t_0 + \dfrac{A^{\frac{1}{c_1+1}} - a_1}{a_2}$
$> n$	$I_1 + I_2 + \cdots + I_{n-1} + I_n < 1$	$t_R = t_0 + t_n + t_0 k_{max}(1 - I_1 - I_2 - ... - I_n)$

Now, following the approach presented in Section 3.4.1, we obtain the expressions listed in Table 3.3 for the retention time of solutes eluted under multilinear ϕ gradient runs performed in different eluent pHs. The various parameters in Table 3.3 are given in the following material for the different equations used for the dependence of $\ln k_0$ on ϕ:

For Equation (3.7),

$$a_1 = c_1\phi_{j-1} - c_1\lambda_j t_{j-1} - c_0, \ a_2 = c_1\lambda_j \tag{3.58}$$

$$A = e^{a_1 + a_2 t_{j-1}} + a_2 D(1 - I_1 - ... - I_{j-1})t_0 \tag{3.59}$$

$$I_j = \frac{[e^{a_1 + a_2 t_j} - e^{a_1 + a_2 t_{j-1}}]}{t_0 a_2 D} \tag{3.60}$$

For Equation (3.8),

$$a_1 = 1 + c_2[\phi_{j-1} - \lambda_j t_{j-1}], \ a_2 = c_2\lambda_j \tag{3.61}$$

For Equation (3.9),

$$a_1 = \phi_{j-1} - \lambda_j t_{j-1}, \ a_2 = \lambda_j \tag{3.62}$$

For both Equations (3.8) and (3.9),

$$A = a_2(c_1 + 1)e^{c_0} D(1 - I_1 - \ldots - I_{j-1})t_0 + \{a_1 + a_2 t_{j-1}\}^{c_1+1} \tag{3.63}$$

$$I_j = \frac{1}{t_0 e^{c_0} D} \frac{\{a_1 + a_2 t_j\}^{c_1+1} - \{a_1 + a_2 t_{j-1}\}^{c_1+1}}{a_2(c_1 + 1)} \tag{3.64}$$

The expressions of Table 3.3 have not been completely tested for multilinear gradient elution yet. This issue is currently under study by our group, but until now, the retention behavior of 16 OPA derivatives of amino acids has been investigated under 19 simple monolinear ϕ gradient conditions performed between two different acetonitrile contents, $\phi_{in} = 0.2$ and $\phi_{max} = 0.5$, with different gradient durations and in different eluent pHs (2.00, 2.50, 3.02, 3.40, 5.05, or 7.00). It was found that the accuracy of the predicted gradient retention times was satisfactory; the absolute average error between experimental and calculated t_R values was only 1.5%, whereas the corresponding maximum error was 8.3% [120].

3.4.4 GRADIENTS IN MOBILE PHASE pH

3.4.4.1 pH Gradients at Different ϕ Values

Until now, only single linear pH gradients have been studied, with the work carried out by Kaliszan's group dominant [81–87]. Moreover, in a recent article [71], we reviewed this work and showed that the empirical model of Equation (3.13) performs better than any other.

Multilinear pH gradients in mobile phases with different organic content ϕ can be treated by direct application of the approach described in Section 3.4.2, that is, by using properly the expressions of Table 3.2. Alternatively, we may follow the treatment of Section 3.4.1 to extend the work carried out in Nikitas, Pappa-Louisi, and Zisi [71] as follows:

A pH-gradient profile may be expressed as

$$pH = \begin{cases} pH_{in} & t < t_D + t_{in} = t_1 \\ pH_1 + \lambda_2(t - t_1) & t_1 < t < t_2 \\ \ldots & \ldots \\ pH_{n-1} + \lambda_n(t - t_{n-1}) & t_{n-1} < t < t_n \\ pH_{max} & t > t_n \end{cases} \tag{3.65}$$

where λ_j is the slope of the pH versus t profile at the jth segment. For simplicity, as is described in the case where the multilinear ϕ gradient is studied, all the slopes may be assumed to be different from zero. Now, if we follow the treatment presented in Section 3.4.1 using Equation (3.13) as the retention model, we readily obtain that

TABLE 3.4

Expressions of the Retention Time under Multilinear pH Gradients at Different ϕ Values When the Retention Model Is Given by Equations (3.13) and (3.66)

Section	Condition	t_R
1	$I_1 \geq 1$	$t_R = t_0(1 + k(pH_{in}))$
j	$I_1 + I_2 + \cdots + I_{j-1} < 1$ $I_1 + I_2 + \cdots + I_{j-1} + I_j \geq 1$	$t_R = t_0 + \dfrac{b\sqrt{1 - b^2 + \left(A_s\lambda_j + e\right)} - b^2 e - A_s\lambda_j}{\left(b^2 - 1\right)\lambda_j}$
$> n$	$I_1 + I_2 + \cdots + I_{n-1} + I_n < 1$	$t_R = t_0 + t_n + \left(1 - I_1 - \ldots - I_n\right)t_0 k(pH_{max})$

the retention time of an analyte under a pH gradient is given by the expressions of Table 3.4, where

$$I_1 = \frac{t_{in} + t_D}{t_0 k(pH_{in})}$$

$$I_j = \frac{a}{t_0}\left[t_j - t_{j-1} + \frac{b}{\lambda_j}\left(\sqrt{1 + (\lambda_j t_j + e)^2} - \sqrt{1 + (\lambda_j t_{j-1} + e)^2} \right) \right], j > 1$$

$$e = pH_{j-1} - \lambda_j t_{j-1} - c$$

$$A_s = t_{j-1} + \left(1 - I_1 - \ldots - I_{j-1}\right)\frac{t_0}{a} + \frac{b}{\lambda_j}\sqrt{1 + (\lambda_j t_{j-1} + e)^2}$$

The analytical expressions for t_R presented in Table 3.4 hold when the modifier content ϕ is kept constant during the pH gradient. Therefore, these expressions can be used for prediction and optimization in pH gradient runs performed in mobile phases with different organic modifier content ϕ, provided that the dependence of a, b, and c on ϕ in Equation (3.13) is known. If the range of ϕ values is relatively narrow, for simplicity we may use the expressions adopted in Nikitas, Pappa-Louisi, and Zisi [71], that is,

$$a = \exp(a_0 + a_1\phi), b = b_0 + b_1\phi \tag{3.66}$$

keeping c constant.

The expressions of Table 3.4 have been tested in a preliminary study for multilinear pH gradients performed in mobile phases containing different values of organic modifier. In more detail, 12 pH gradients from 3.2 to 9.0 were performed by applying a monolinear pump program between two mobile phases, which consisted of aqueous phosphate buffers with the different pH values mentioned and a fixed concentration of acetonitrile as the organic modifier ($\phi = 0.25, 0.27, 0.3,$ or 0.35). However, the actual pH changes were curved with a shape depicted in Figure 3.13 for a gradient duration of

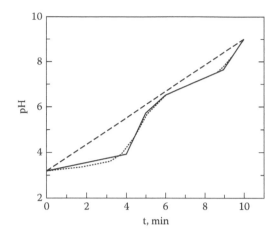

FIGURE 3.13 Applied monolinear pump pH program (- - -) and the actual curved pH profile (· · ·) approximated by five linear segments (–).

10 min. For this reason, the actual pH profiles mentioned should be treated as multi-linear gradients. The retention behavior of 17 OPA derivatives of amino acids obtained under the pH gradient profiles described with different gradient durations was very well predicted using the expressions of Table 3.4 since it was found that the absolute average error between all experimental and calculated t_R values was only 1.5%, whereas the corresponding maximum error was 6.3% (unpublished results).

3.4.4.2 Combined pH/ϕ Gradients

Combined multilinear pH/ϕ gradients can be straightforwardly treated following the approach developed in Section 3.4.2. The retention model that may be used is a combination of Equations (3.7) and (3.12), which is

$$k(pH, \phi) = \frac{k_{00} e^{k_{01}\phi} + k_{10} e^{k_{11}\phi} 10^{i(pH - pK)}}{1 + 10^{i(pH - pK)}} \tag{3.67}$$

Such a study is currently being carried out by our group. The results about the performance of this approach are now available [121].

3.4.5 Gradients in Flow Rate

Flow programming is rather rarely used; this gradient mode is sometimes combined with solvent gradient elution or with temperature programming. When only the flow rate F varies, the retention factor k is constant, and Equation (3.5) yields [2,18,19,45,54]

$$\frac{1}{t_{01}(1+k)} \int_0^{t_R} F(t) \, dt = 1 \tag{3.68}$$

where t_{01} is the column holdup time that corresponds to $F = 1$ in arbitrary units. This equation is easily solved with respect to t_R, especially when a linear or multilinear gradient is used.

Suppose a multilinear gradient of F versus t, like that in Figure 3.1 with F in place of ϕ and $t_D = 0$. Thus, in the jth linear segment ($j > 1$), that is, between t_{j-1} and t_j, the flow rate varies according to $F = F_{j-1} + \lambda_j(t - t_{j-1})$. When the analyte is eluted in the jth linear segment, that is, between t_{j-1} and t_j, the condition is again given by Equations (3.29), where I_j is now given by

$$I_j = \frac{1}{t_{01}(1+k)} \int_{t_{j-1}}^{t_j} F dt = \frac{F_j + F_{j-1}}{2t_{01}(1+k)}(t_j - t_{j-1}), j = 1, 2, \ldots \tag{3.69}$$

The retention time t_R of the analyte may be calculated from

$$\int_{t_{j-1}}^{t_R} F dt = t_{01}(1+k)(1 - I_1 - I_2 - \cdots - I_{j-1}) \tag{3.70}$$

which results in (Table 3.5)

$$t_R = \frac{-F_{0j} + \sqrt{F_{0j}^2 + 2\lambda_j C_j}}{\lambda_j} \tag{3.71}$$

TABLE 3.5

Expressions of the Retention Time at the jth Section of a Multilinear F- and T-Gradient Profile When the Retention Model for T Gradients Is Given by Equation (3.75)

Condition	t_R
$I_1 + I_2 + \cdots + I_{j-1} < 1$	I. Multilinear F gradient
$I_1 + I_2 + \cdots + I_{j-1} + I_j \geq 1$	
	$\lambda_j \neq 0:\ t_R = \dfrac{-F_{0j} + \sqrt{F_{0j}^2 + 2\lambda_j C_j}}{\lambda_j}$
	$\lambda_j = 0:\ t_R = t_{01}(1+k)(1 - I_1 - I_2 - \cdots - I_{j-1})/F_{j-1} + t_{j-1}$
	II. Multilinear T gradient
	$\lambda_j \neq 0:\ t_R = \dfrac{1}{a_{1j}}\ln(e^{B_j} - e^{a_{0j}})$
	$\lambda_j = 0:\ t_R = t_{j-1} + t_0 S_{j-1}(1 + e^{a_{0j}})$

where

$$F_{0j} = F_{j-1} - \lambda_j t_{j-1} \qquad (3.72)$$

and

$$C_j = t_{01}(1+k)(1-I_1-I_2-\cdots-I_{j-1}) + \lambda_j t_{j-1}^2/2 + F_{0j}t_{j-1} \qquad (3.73)$$

If $\lambda_j = 0$, then Equation (3.70) yields

$$t_R = t_{01}(1+k)(1-I_1-I_2-\cdots-I_{j-1})/F_{j-1} + t_{j-1} \qquad (3.74)$$

These equations have been tested [45]; it was shown that a simple linear flow rate gradient can hardly be used in optimization separation. A substantial optimizing effect can be achieved on a separation by multilinear or more complicated flow rate gradients, where the applying variation in mobile phase flow rate can change peak retention both absolutely and relatively, yielding an optimum rearrangement of peaks within the chromatogram. Figure 3.14b shows such a chromatogram in which the overall separation of nine underivatized amino acids, beta-(3,4-dihydroxyphenyl)-l-alanine (dopa), l-tyrosine (tyr), l-a-methyl-dopa (me-dopa), dl-mtyrosine (m-tyr), dl-alpha-methyltyrosine (me-tyr), 5-hydroxytryptophan (5htp), l-phenylanine (phe), 3-nitro-l-tyrosine (ntyr), and l-tryptophan (trp) in aqueous phosphate buffers (pH 2.5) modified with acetonitrile, was significantly improved even compared to the lowest constant flow rate of 0.5 mL/min separation depicted in Figure 3.14a. The chromatograms were taken from Pappa-Louisi, Nikitas, and Zitrou [45]. The separation capability of this flow rate plot is given by the mild starting decrease of the flow rate, followed by a sharp increase of the flow rate at the correct time point, which after a certain time length is kept constant to the end of the run.

3.4.6 GRADIENTS IN COLUMN TEMPERATURE

Temperature gradients, like flow rate gradients, act on the column immediately as they are programmed. Thus, we have $t_D = 0$. The theory of linear and multilinear temperature gradients was developed by Snyder's group two decades ago [3,54,122,123]. Initially, the theory was developed for gas chromatography, but it is directly applicable to liquid chromatography since the fundamental Equation (3.5) is the same for both gas and liquid chromatography.

The theory was applied to liquid chromatography recently by Schmidt's group [124–128] possibly due to the appearance of hysteresis phenomena during T gradients [19,20,129]. In fact, a hysteresis appears between the actual temperature of the oven and the effective temperature inside the column when we use conventional chromatographic columns. Apart from this hysteresis, there may be a lag between the programmed and the actual temperature formed in the oven if the oven does not respond fast in the programmed changes of the temperature [129]. The hysteresis between the oven temperature and the effective column temperature can

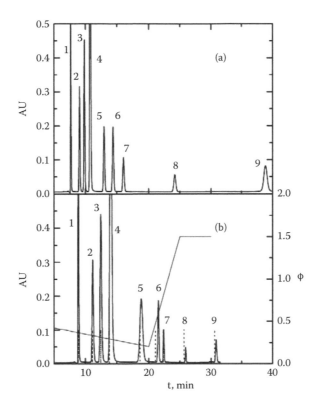

FIGURE 3.14 UV chromatograms of a mixture of underivatized amino acids obtained at a constant flow rate of 0.5 mL/min (a) and by flow rate program indicated by the segmented line (b); the volume fraction of acetonitrile in the mobile phase is $\phi = 0.06$. Peaks are 1, dopa; 2, tyr; 3, me-dopa; 4, m-tyr; 5, me-tyr; 6, 5htp; 7, phe; 8, n-tyr; and 9, trp. The dotted vertical lines indicate the predicted retention times by means of Equation (3.71). (Reprinted from A. Pappa-Louisi, P. Nikitas, and A. Zitrou, *Anal. Chim. Acta*, 573–574: 305, 2006, with permission.)

be overcome by using micro- or capillary separation columns. The lag between programmed and oven temperature can be minimized by using proper heating systems. Such a system was described in Wiese, Teutenberg, and Schmidt [124,125] and consists of three modules: the eluent preheating unit, the column heating unit, and the eluent cooling unit. The heat transfer is achieved by block heating, which means that the capillaries and column are tightly enclosed by aluminum blocks.

Note that the fundamental Equation (3.5) does not have an analytical solution with respect to t_R when we use the expressions for k given by Equations (3.14) and (3.15). For this reason, Snyder et al. [122,123] developed the linear elution strength (LES) model, which is based on the linear dependence of ln k on T:

$$\ln k = c_0 - c_1 T \tag{3.75}$$

When Equation (3.75) is introduced into Equation (3.5), we obtain

$$t_R = \frac{t_0}{b_T}\ln[e^{b_T}(k_0+1)-k_0] \tag{3.76}$$

with

$$b_T = t_0 c_1 \Delta T / t_G \tag{3.77}$$

where t_0 is the column dead time, k_0 is the retention factor of the solute at the start of the temperature gradient, $\Delta T = T_{final} - T_{start}$, and t_G is the temperature gradient time.

For segmented temperature gradients, Equation (3.76) is extended to [3,54,122,123]

$$t_R = \frac{t_0}{b_T}\ln[e^{rb_T}(k_0+1)-k_0] \tag{3.78}$$

when $b_T \neq 0$ and

$$t_R = rt_0(k_0+1) \tag{3.79}$$

if $b_T = 0$. Here, r is the distance in longitudinal direction that an analyte moves through the column during a temperature segment. Therefore, the sum of the fractional migration r of an analyte across the column during each temperature segment is equal to one.

Note that Equation (3.78) or Equation (3.79) describes the change of the retention factor of an analyte during each temperature segment, and its application requires the retention factor k_r of the analyte at the end of a temperature segment. This value then has to be used as the initial value instead of k_0 for the next temperature segment. k_r may be calculated from

$$\ln k_r = \ln k_0 - b_T t_R / t_0 \tag{3.80}$$

Finally, the sum of the calculated retention times of an analyte for all temperature segments represents the total retention time of a multilinear temperature gradient.

The multilinear temperature gradient elution theory has been recently employed for systematic method development in high-temperature high-performance liquid chromatography (HT-HPLC) [125,128]. In Wiese, Teutenberg, and Schmidt [125], it was found that an average relative error of predicted retention times of 2.7% and 1.9% was observed for simulations based on isothermal and temperature gradient measurements, respectively. In addition, it was shown that the accuracy of retention time predictions could be improved to around 1.5% if c_1 were calculated as temperature dependent. The theory was also applied to the separation of a mixture of food additives, including theobromine, theophylline, catechine, caffeine, aspartame, rutin, and uracil. In this case, the average relative error of predicted retention times of complex segmented temperature gradients was less than 5%. Finally, multilinear temperature gradient elution was applied for the separation of five sulfonamides and uracil [126]; a typical chromatogram is shown in Figure 3.15, where we see a baseline separation of the mixture of analytes.

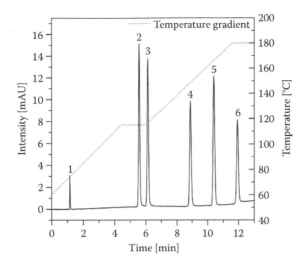

FIGURE 3.15 Separation of five sulfonamides and uracil by multilinear temperature gradient elution. Stationary phase: Waters XBridge C18 (75 × 4.6 mm, 2.5 μm); mobile phase: deionized water with 0.1% formic acid; flow rate: 1.0 mL/min; injection volume: 1 μL; detection: UV at 270 nm. Analytes: 1, uracil; 2, sulfadiazine; 3, sulfathiazole; 4, sulfamerazine; 5, sulfamethoxazole; and 6, sulfamethazine. (Reprinted from S. Wiese, Th. Teutenberg, and T. Schmidt, *J. Chromatogr. A*, 1218: 6898, 2011, with permission.)

Alternatively, the theory of multilinear temperature gradient elution may be formulated within the frames of the treatment presented throughout this chapter, in particular in Section 3.4.1. Thus, in the jth linear segment, that is, between t_{j-1} and t_j, the temperature is given by $T = T_{j-1} + \lambda_j(t - t_{j-1})$, and $\ln k$ may be expressed as

$$\ln k = a_{0j} - a_{1j}t \tag{3.81}$$

where

$$a_{0j} = c_0 - c_1 T_{j-1} + c_1 \lambda_j t_{j-1} \text{ and } a_{1j} = c_1 \lambda_j \tag{3.82}$$

The condition for the analyte to be eluted in the jth linear segment may be expressed again from Equations (3.29), where I_j is calculated from

$$I_j = \int_{t_{j-1}}^{t_j} \frac{dt}{t_0(1+k)} = \frac{1}{t_0 a_{1j}} \ln \frac{1 + 1/k_j}{1 + 1/k_{j-1}} \tag{3.83}$$

In this case, the retention time t_R of the analyte is calculated from

$$\int_{t_{j-1}}^{t_R} \frac{dt}{t_0 k} = 1 - \int_0^{t_{j-1}} \frac{dt}{t_0 k} = 1 - I_1 - I_2 - \cdots - I_{j-1} = S_{j-1} \tag{3.84}$$

which results in (Table 3.5)

$$t_R = \frac{1}{a_{1j}} \ln(e^{B_j} - e^{a_{0j}}) \tag{3.85}$$

where

$$B_j = t_0 a_{1j} S_{j-1} + \ln(e^{a_{0j}} + e^{a_{1j} t_{j-1}}) \tag{3.86}$$

If $\lambda_j = 0$, then

$$I_j = \frac{1}{t_0} \frac{t_j - t_{j-1}}{1 + e^{a_{0j}}} \tag{3.87}$$

and

$$t_R = t_{j-1} + t_0 S_{j-1}(1 + e^{a_{0j}}) \tag{3.88}$$

Finally, analytical solutions with respect to t_R may be obtained for every retention model by approximating the multilinear gradient or every other gradient by a step-wise one. Then, t_R is estimated from [128]

$$t_R = t_0(1 + k_{T_n}) + \delta t \frac{k_{T_1} - k_{T_n}}{1 + k_{T_1}} + \delta t \frac{k_{T_2} - k_{T_n}}{1 + k_{T_2}} + \cdots + \delta t \frac{k_{T_{n-1}} - k_{T_n}}{1 + k_{T_{n-1}}} \tag{3.89}$$

where δt is the time step of the stepwise gradient, $k_{T_j} = (t_{T_j} - t_0)/t_0$, t_{T_j} is the retention time of the analyte if the temperature were constant and equal to that of the jth step, and n is the least number of terms of the sum that makes the following inequality valid:

$$\frac{\delta t}{t_0(1 + k_{T_1})} + \frac{\delta t}{t_0(1 + k_{T_2})} + \cdots + \frac{\delta t}{t_0(1 + k_{T_n})} \geq 1 \tag{3.90}$$

This theory has been tested using six alkylbenzenes—benzene, toluene, eth-ylbenzene, isopropylbenzene, propylbenzene, and *tert*-butylbenzene—in eluting systems modified by acetonitrile [128]. The chromatographic column was a con-ventional Zorbax SB-C18 column (3.5 μm, 150 × 4.6 mm), stable at temperatures of 90°C or less. The initial temperature in all temperature gradients was 15°C, and the final was 75°C. We found that the retention prediction was excellent, provided that hysteresis phenomena were properly embodied in the treatment. In this case, the average percentage error between experimental and predicted retention times was less than 2%.

3.5 CONCLUSIONS

The work presented in this chapter continues our attempts to computer-assisted devel-opments of effective multilinear gradient elution optimization procedures. For this purpose, the analytical expressions for the retention of solutes obtained under different single- or multimode multilinear gradient conditions were derived based on several models considering the main factors affecting retention, i.e., organic modifiers, pH (for ionogenic analytes), temperature, and flow rate. Moreover, the derived mathematical equations were clearly presented in separate tables for each kind of multilinear gradient mode to encourage chromatographers to apply them in analyses of complex mixtures. However, a full utilization of multilinear gradient elution needs dedicated optimization software, which can be obtained with instructions by e-mailing us.

ACKNOWLEDGMENT

This research was cofinanced by the European Union (European Social Fund, ESF) and Greek national funds through the Operational Program "Education and Lifelong Learning" of the National Strategic Reference Framework (NSRF)—Research Funding Program: Heracleitus II. Investing in knowledge society through the European Social Fund.

REFERENCES

1. L. R. Snyder and J. J. Kirkland, *Introduction to Modern Liquid Chromatography*, 2nd ed., Wiley-Interscience, New York, 1979.
2. P. Jandera and J. Churacek, *Gradient Elution in Column Liquid Chromatography*, Elsevier, Amsterdam, 1985.
3. L. R. Snyder, J. L. Glajch, and J. J. Kirkland, *Practical HPLC Method Development*, Wiley-Interscience, New York, Chapter 8, 1997.
4. C. F. Poole, *The Essence of Chromatography*, Elsevier, Amsterdam, 2003.
5. H. Schmidt-Traub, *Preparative Chromatography of Fine Chemical and Pharmaceutical Agents*, Wiley-VCH, New York, 2005.
6. L. R. Snyder and J. W. Dolan, *High Performance Gradient Elution*, Wiley-Interscience, New York, 2007.
7. E. C. Freiling, *J. Am. Chem. Soc.*, 77: 2067 (1955).
8. E. C. Freiling, *J. Phys. Chem.*, 61: 543 (1957).
9. B. Drake, *Akriv. Kemi.*, 8: 1 (1955).
10. L. R. Snyder, *J. Chromatogr.*, 13: 415 (1964).
11. L. R. Snyder, *Chromatogr. Rev.*, 7: 1 (1965).
12. L. R. Snyder and D. L. Saunders, *J. Chromatogr. Sci.*, 7: 145 (1969).
13. L. R. Snyder, in *High Performance Liquid Chromatography*, Horvath, C., Ed., Academic Press: New York, Vol. 1: p. 207, 1980.
14. M. A. Quarry, R. L. Grob, and L. R. Snyder, *Anal. Chem.*, 58: 907 (1986).
15. P. Nikitas and A. Pappa-Louisi, *Anal. Chem.*, 77: 5670 (2005).
16. W. Q. Hao, X. M. Zhang, and K. Y. Hou, *Anal. Chem.* 78: 7828 (2006).
17. P. Nikitas, A. Pappa-Louisi, and P. Balkatzopoulou, *Anal. Chem.*, 78: 5774 (2006).

18. A. Pappa-Louisi, P. Nikitas, P. Balkatzopoulou, and G. Louizis, *Anal. Chem.,* 79: 3888 (2007).
19. P. Nikitas, A. Pappa-Louisi, K. Papachristos, and C. Zisi, *Anal. Chem.,* 80: 5508 (2008).
20. A. Pappa-Louisi, P. Nikitas, K. Papachristos, and P. Balkatzopoulou, *Anal. Chem.,* 81: 1217 (2009).
21. P. Nikitas and A. Pappa-Louisi, *J. Liq. Chromatogr. Relat. Technol.,* 32: 1527 (2009).
22. A. Pappa-Louisi, P. Agrafiotou, and S. Sotiropoulos, *Methods Mol. Biol.,* 828: 101 (2012).
23. I. Nistor, M. Cao, B. Debrus, P. Lebrun, F. Lecomte, E. Rozet, L. Angenot, M. Frederich, R. Oprean, and P. Hubert, *J. Pharm. Biomed. Anal.,* 56: 30 (2011).
24. P. Nikitas, A. Pappa-Louisi, P. Agrafiotou, and A. Mansour, *J. Chromatogr. A,* 1218: 5658 (2011).
25. A. Susanto, K. Treier, E. Knieps-Grünhagen, E. Von Lieres, and J. Hubbuch, *Chem. Eng. Technol.,* 32: 140 (2009).
26. K. Kamel and M. R. Hadjmohammadi, *Chromatographia,* 67: 169 (2008).
27. J. W. Lee and K. H. Row, *J. Liq. Chromatogr. Relat. Technol.,* 31: 2401 (2008).
28. P. Nikitas, A. Pappa-Louisi, and A. Papageorgiou, *J. Chromatogr. A* 1157: 178 (2007).
29. V. Concha-Herrera, J. R. Torres-Lapasió, G. Vivó-Truyols, and M. C. García-Álvarez-Coque, *Anal. Chim. Acta,* 582: 250 (2007).
30. P. Nikitas, A. Pappa-Louisi, and P. Agrafiotou, *J. Chromatogr. A,* 1120: 299 (2006).
31. T. Jupille, L. Snyder, and I. Molnar, *LC-GC Eur.,* 15: 596 (2002).
32. M. F. Tutunji and Q. M. Al-Mahasneh, *J. Toxicol. Clin. Toxicol.,* 32: 267 (1994).
33. U. Turpeinen and U.-M. Pomoell, *Clin. Chem.,* 31: 1710 (1985).
34. M. R. Hadjmohammadi and K. Kamel, *J. Iran. Chem. Soc.,* 7: 107 (2010).
35. K. A. Georga, V. F. Samanidou, and I. N. Papadoyannis, *J. Liq. Chromatogr. Relat. Technol.,* 23: 1523 (2000).
36. K. A. Georga and V. F. Samanidou, *J. Liq. Chromatogr. Relat. Technol.,* 22: 2975 (1999).
37. I. N. Papadoyannis, V. F. Samanidou, and K. A. Georga, *J. Liq. Chromatogr. Relat. Technol.,* 19: 2559 (1996).
38. P. Chaminade, A. Baillet, and D. Ferrier, *J. Chromatogr. A,* 672: 67 (1994).
39. M. Beneito-Cambra, V. Bernabe-Zafon, J. M. Herrero-Martinez, E. F. Simo-Alfonso, and G. Ramis-Ramos, *Talanta,* 79: 275 (2009).
40. Y. Shan, W. Zhang, A. Seidel-Morgenstern, R. Zhao, and Y. Zhang, *Sci. China Ser. B,* 49: 315 (2006).
41. X.-Q. Ma, L.-X. Wang, Q. Xu, F. Zhang, H.-B. Xiao, and X.-M. Liang, *Chem. J. Chinese U.,* 25: 238 (2004).
42. H. Xiao, X. Liang, and P. Lu, *J. Sep. Sci.,* 24: 186 (2001).
43. A. Pappa-Louisi, P. Agrafiotou, and S. Sotiropoulos, *Curr. Org. Chem.,* 14: 2235 (2010).
44. A. Pappa-Louisi, P. Nikitas, and A. Papageorgiou, *J. Chromatogr. A,* 1166: 126 (2007).
45. A. Pappa-Louisi, P. Nikitas, and A. Zitrou, *Anal. Chim. Acta,* 573–574:305 (2006).
46. J. García-Lavandeira, P. Oliveri, J. A. Martínez-Pontevedra, M. H. Bollaín, M. Forina, and R. Cela, *Anal. Bioanal. Chem.,* 399: 1951 (2011).
47. R. Kaliszan and P. Wiczling, *TrAC Trends Anal. Chem.,* 30: 1372 (2011).
48. V. Concha-Herrera, G. Vivo-Truyols, J. R. Torres-Lapasio, and M. C. Garcia-Alvarez-Coque, *J. Chromatogr.,* 1063: 79 (2005).
49. H. Gika, G. Theodoridis, F. Mattivi, U. Vrhovsek, and A. Pappa-Louisi, *J. Sep. Sci.,* 35: 376–383 (2012).
50. H. Gika, G. Theodoridis, F. Mattivi, U. Vrhovsek, and A. Pappa-Louisi, *Anal. Bioanal. Chem.,* 404: 701 (2012).
51. A. Pappa-Louisi, P. Agrafiotou, D. Thomas, and K. Papachristos, *Chromatographia,* 71: 571 (2010).
52. A. Pappa-Louisi and Ch. Zisi, *Talanta,* 93: 279 (2012).

53. A. Pappa-Louisi, P. Agrafiotou, and K. Papachristos, *Anal. Bioanal. Chem.*, 397: 2151 (2010).
54. L. R. Snyder and J. W. Dolan, *Adv. Chromatogr.*, 38: 115 (1998).
55. P. J. Schoenmakers, H. A. H. Billiet, R. Tijssen, and L. De Galan, *J. Chromatogr.*,149: 519 (1978).
56. P. Jandera and J. Churacek, *Adv. Chromatogr.*, 19: 125 (1980).
57. P. Jandera, *Adv. Chromatogr.*, 43: 1 (2005).
58. K. Valko, L. R. Snyder, and J. L. Glajch, *J. Chromatogr. A*, 656: 501 (1993).
59. P. J. Schoenmakers, H. A. H. Biliiet, and L. de Galan, *J. Chromatogr.*, 282: 107 (1983).
60. P. J. Schoenmakers, H. A. H. Billiet, and L. de Galan, *Chromatographia*, 15: 205 (1982).
61. P. Nikitas, A. Pappa-Louisi, and P. Agrafiotou, *J. Chromatogr. A*, 946: 33 (2002).
62. A. Pappa-Louisi, P. Nikitas, P. Balkatzopoulou, and C. Malliakas, *J. Chromatogr. A*, 1033:29 (2004).
63. U. D. Neue, C. H. Phoebe, K. Tran, Y.-F. Cheng, and Z. J. Lu,. *J. Chromatogr. A*, 925: 49 (2001).
64. Cs. Horvath, W. Melander, and I. Molnar, *Anal. Chem.*, 49: 142 (1977).
65. R. M. Lopes-Marques and P. J. Schoenmakers, *J. Chromatogr.*, 592: 157 (1992).
66. P. J. Schoenmakers and R. Tijssen, *J. Chromatogr. A*, 656: 577 (1993).
67. M. Roses, I. Canals, H. Allemann, K. Siigur, and E. Bosch, *Anal. Chem.*, 68: 4094 (1996).
68. J. E. Hardcastle and I. Jano, *J. Chromatogr. B*, 717: 39 (1998).
69. M. Roses and E. Bosch, *J. Chromatogr. A*, 982: 1 (2002).
70. P. Nikitas and A. Pappa-Louisi, *J. Chromatogr. A*, 971: 47 (2002).
71 P. Nikitas, A. Pappa-Louisi, and Ch. Zisi, *Anal. Chem.*, 84: 6611 (2012).
72. P. L. Zhu, L. R. Snyder, J. W. Dolan, N. M. Djordjevic, D. W. Hill, L. C. Sander, and T. J. Waeghe, *J. Chromatogr. A*, 756: 21 (1996).
73. A. Pappa-Louisi, P. Nikitas, K. Papachristos, and C. Zisi, *J. Chromatogr. A*, 1201, 27 (2008).
74. R. G. Wolcott, J. W. Dolan, and L. R. Snyder, *J. Chromatogr. A*, 869: 3 (2000).
75. J. W. Dolan, L. R. Snyder, N. M. Djordjevic, D. W. Hill, D. L. Saunders, L. Van Heukelem, and T. J. Waeghe, *J. Chromatogr. A*, 803: 1 (1998).
76. L. R. Snyder and J. W. Dolan, *J. Chromatogr. A*, 892: 107 (2000).
77. J. W. Dolan, L. R. Snyder, R. G. Wolcott, P. Haber, T. Baczek, R. Kaliszan, and L. C. Sander, *J. Chromatogr. A*, 857: 41 (1999).
78. S. Heinisch, G. Puy, M.-P. Barrioulet, and J.-L. Rocca, *J. Chromatogr. A*, 1118: 234 (2006).
79. K. Jinno and M. Yamagami, *Chromatographia*, 27: 417 (1989).
80. P. Nikitas and A. Pappa-Louisi, *J. Chromatogr. A*, 1216: 1737 (2009).
81. R. Kaliszan, P. Wiczling, and M. J. Markuszewski, *Anal. Chem.*, 76: 749 (2004).
82. P. Wiczling, M. J. Markuszewski, and R. Kaliszan, *Anal. Chem.*, 76: 3069 (2004).
83. R. Kaliszan, P. Wiczling, and M. J. Markuszewski, *J. Chromatogr. A*, 1060: 165 (2004).
84. R. Kaliszan and P. Wiczling, *Anal. Bioanal. Chem.*, 382: 718 (2005).
85. P. Wiczling, P. Kawczak, A. Nasal, and R. Kaliszan, *Anal. Chem.*, 78: 239 (2006).
86. P. Wiczling and R. Kaliszan, *Anal. Chem.*, 82: 3692 (2010).
87. P. Wiczling and R. Kaliszan, *J. Chromatogr. A*, 1217: 3375 (2010).
88. A. Tellez, M. Roses, and E. Bosch, *Anal. Chem.*, 81: 9135 (2009).
89. A, Pappa-Louisi, P. Nikitas, and P Agrafiotou, *J. Chromatogr. A*, 1127: 21 (2005).
90. A. Pappa-Louisi and P. Nikitas, *Chromatographia*, 57: 169 (2003).
91. N. R. Draper and H. Smith, *Applied Regression Analysis*, Wiley, New York, 1981.
92. K. Levenberg, *Q. Appl. Math.*, 2: 164 (1944).
93. D. Marquardt and J. Siam, *Appl. Math.*, 11: 431 (1963).
94. W. H. Press, S. A. Teukolsky, W. T. Vetterling, and B. P. Flannery, *Numerical Recipes in C*, 2nd ed., Cambridge University Press, Cambridge, UK, 1992.
95. P. Nikitas and A. Pappa-Louisi, *Chromatographia*, 52: 477 (2000).
96. R. de Levie, *J. Chem. Educ.*, 76: 1594 (1999).

97. R. De Levie, *Excel in Analytical Chemistry*, Cambridge University Press, Cambridge, UK, 2001.
98. C. Comuzzi, P. Polese, A. Melchior, R. Portanova, and M. Tolazzi, *Talanta*, 59: 67 (2003).
99. Z. Michalewicz, *Genetic Algorithms + Data Structures = Evolution Programs,* Springer, Berlin, 1996.
100. D. E. Goldberg, *Genetic Algorithms in Search, Optimization and Machine Learning*, Addison-Wesley, Reading, MA, 1989.
101. D. T. Pham and D. Karaboga, *Intelligent Optimization Techniques*, Springer, Berlin, 2000.
102. P. Nikitas, A. Pappa-Louisi, A. Papageorgiou, and A. Zitrou, *J. Chromatogr. A*, 942: 93 (2002).
103. P. Nikitas and A. Papageorgiou, *Comput. Phys. Commun.*, 141: 225 (2001).
104. J. C. Berridge, *Techniques for the Automated Optimization of HPLC Separations*; Wiley, Chichester, UK, 1985.
105. R. Cela, J. A. Martinez, C. Gonzalez-Barreiro, and M. Lores, *Chemometr. Intell. Lab. Syst.*, 69: 137 (2003).
106. A. K. Smilde, A. Knevelman, and P. M. J. Coenegracht, *J. Chromatogr.*, 369: 1 (1986).
107. H. R. Keller and D. L. Massart, *Trends Anal. Chem.*, 9: 251 (1990).
108. M. J. Sáiz-Abajo, J. M. González-Sáiz, and C. Pizarro, *Anal. Chim. Acta*, 528: 63 (2005).
109. M. R. Hadjmohammadi and F. J. Safa, *Sep. Sci.*, 27: 997 (2004).
110. G. Derringer and R. Suich, *J. Qual. Technol.*, 12: 214 (1980).
111. B. Bourguignon and D. L. Massart, *J. Chromatogr.*, 586: 1 (1991).
112. F. Safa and M. R. Hadjmohammadi, *J. Chromatogr. A*, 1078: 42 (2005).
113. J. L. Glajch and J. J. Kirkland, *J. Chromatogr.*, 485: 51 (1989).
114. C. Ortiz-Bolsico, J. R. Torres-Lapasió, and M. C. García-Álvarez-Coque, *J. Chromatogr. A*, 1229: 180 (2012).
115. A. Ortín, J. R. Torres-Lapasió, and M. C. García-Álvarez-Coque, *J. Chromatogr. A*, 1218: 5829 (2011).
116. P. Nikitas and A. Pappa-Louisi, *J. Chromatogr. A*, 1068: 279 (2005).
117. A. Pappa-Louisi, P. Nikitas, and M. Karageorgaki, *J. Chromatogr. A*, 1091: 97 (2006).
118. P. J. Schoenmakers, H. A. H. Billiet, and L. D. Galan, *J. Chromatogr.*, 218: 261 (1981).
119. A. Andres, A. Tellez, M. Roses, and E. Bosch, *J. Chromatogr. A*, 1247: 71 (2012).
120. S. Fasoula, Ch. Zisi, P. Nikitas, and A. Pappa-Louisi, *J. Chromatogr. A*, 1305: 131 (2013).
121. Ch. Zisi, S. Fasoula, A. Pappa-Louisi, and P. Nikitas, *Anal. Chem.*, 85: 9514 (2013).
122. D. E. Bautz, J. W. Dolan, W. D. Raddatz, and L. R. Snyder, *Anal. Chem.*, 62: 1560 (1990).
123. D. E. Bautz, J. W. Dolan, and L. R. Snyder, *J. Chromatogr.*, 541: 1 (1991).
124. S. Wiese, Th. Teutenberg, and T. Schmidt, *Anal. Chem.*, 83: 2227 (2011).
125. S. Wiese, Th. Teutenberg, and T. Schmidt, *J. Chromatogr. A*, 1222: 71 (2012).
126. S. Wiese, Th. Teutenberg, and T. Schmidt, *J. Chromatogr. A*, 1218: 6898 (2011).
127. J. Haun, Th. Teutenberg, and T. Schmidt, *J. Sep. Sci.*, 35: 1723 (2012).
128. L. Zhang, D. Kujawinski, M. Jochmann, and T. Schmidt, *Rapid Commun. Mass Spectrom.*, 25: 2971 (2011).
129. A. Pappa-Louisi, P. Nikitas, Chr. Zisi, and K. Papachristos, *J. Sep. Sci.*, 31: 2953 (2008).

4 Analytical Separation of Enantiomers by Gas Chromatography on Chiral Stationary Phases

Volker Schurig and Markus Juza

4.1   INTRODUCTION

Enantioselective gas chromatography (enantio-GC) on chiral stationary phases (CSPs) constitutes an important tool in the determination of enantiomeric compositions in different areas of contemporary research (Figure 4.1) [1].

The recent state of the art is treated in general surveys [2–4]. Topical reviews focus either on the use of chiral metal complexes [5] and modified cyclodextrins (CDs) as CSPs [6] or on selected enantioseparations of important classes of chiral compounds such as derivatized α-amino acids [7], unfunctionalized saturated hydrocarbons [8], inhalational anesthetics [9], and environmental organochlorines [10–12].

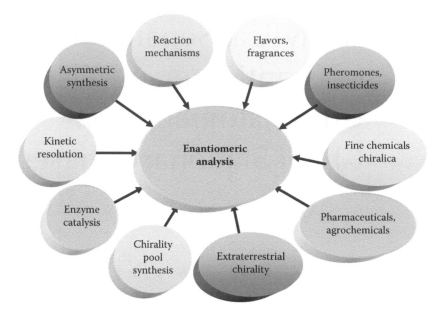

FIGURE 4.1   Precise enantiomeric analyses are required in many different fields of research and applications.

Whereas analytical enantioseparations by enantio-GC are widespread, (semi)preparative enantioseparations on CSPs are less important [13].

The separation of enantiomers by enantio-GC requires the presence of a nonracemic chiral auxiliary acting as a resolving agent in the spirit of Pasteur's resolution principles via diastereomeric relationships. In regard to the use of the nonracemic resolving agent, an indirect and a direct approach have been employed. (a) By the *indirect* approach, diastereomeric derivatives are formed off column via the reaction of the enantiomers with the nonracemic chiral auxiliary. The diastereomers are then separated on a conventional achiral stationary phase [14]. This method requires the presence of suitable chemical functionalities enabling a quantitative chemical transformation, the absence of kinetic resolution, the absence of racemization during the derivatization, and an unbiased detection of diastereomers that may differ in their physical properties. The enantiomeric purity of the chiral auxiliary must be known with certainty. The measured diastereomeric excess $de_{measured}$ must be corrected for the enantiomeric excess ee_{aux} of the chiral auxiliary in case it is not enantiomerically pure, that is, $ee_{substrate} = de_{measured}/ee_{aux}$ [15]. (b) By the *direct* approach, derivatized or underivatized enantiomers are separated on column via the fast and reversible noncovalent diastereomeric association with a nonracemic CSP of high but not necessarily total enantiomeric purity. The on-column formation of diastereomeric association complexes of distinct stabilities causes differences in the partition coefficients of the enantiomers, resulting in GC resolution [16]. Only the straightforward direct approach in enantio-GC is discussed here.

High efficiency, sensitivity, and speed of analysis are important advantages of enantioseparations by high-resolution capillary gas chromatography (enantio-HRC-GC). Due to the high separation power of enantio-HRC-GC, contaminants and impurities are separated from the chiral analytes, and the simultaneous analysis of multicomponent mixtures of enantiomers (e.g., all derivatized proteinogenic α-amino acids) can be achieved. An enantioseparation factor α as low as 1.02 leads to a baseline resolution of the enantiomers by enantio-HRC-GC. Hyphenated techniques, such as multidimensional gas chromatography (MD-GC), that is, in series-coupled column operation, interfacing and coupling methods such as gas chromatography and mass spectrometry (enantio-GC-MS) are important tools in chiral analysis (Section 4.6). Employing the selected ion-monitoring mode, trace amounts of enantiomers can be detected by enantio-GC-MS(SIM) (selected ion monitoring). The detection of enantiomers in enantio-GC is straightforward. The universal flame ionization detector (FID) is linear over five orders of magnitude, and detection sensitivity can further be enhanced by electron capture detection (ECD) and by element-specific detection. In contrast to liquid chromatography or electromigration methods, the delicate choice of solvents, buffers, modifiers, and gradient elution systems is absent in GC. Yet, the prerequisites for enantio-GC are volatility, thermal stability, and resolvability of the enantiomers, restricting its general application. The merit of enantio-GC has been recognized from its very beginning, and a number of past reviews are available [16–24]. Chapters on enantio-GC are also included in general treatises on chiral analysis [25–28]. Enantio-GC is performed using three distinct CSPs based on (a) hydrogen bonding, (b) metal ion complexation, and (c) inclusion. They are treated successively here.

4.2 CHIRAL STATIONARY PHASES BASED ON α-AMINO ACID DERIVATIVES FOR HYDROGEN-BONDING ENANTIO-GC

Gil-Av, Feibush, and Charles-Sigler in 1966 described the first reproducible and *direct* enantioseparation of derivatized α amino acids on α-amino-acid-derived CSPs [29]. The design of this enantioselective selector–selectand system was based on the idea of biomimetically imitating the stereoselective peptide–enzyme interaction employing simple α-amino acid entities as model substances. This achievement was the basis for all further developments in enantio-GC [1]. A prerequisite of this pioneering advance consisted of the availability of high-resolution open-tubular columns [30]. Thus, a home-made 100 m × 0.25 mm inside diameter (ID) glass capillary column was coated by the plug method with a 20% solution of the CSP N-trifluoroacetyl-L-isoleucine lauryl ester (Figure 4.2) in diethylether. The 2-propanol, n-butanol, and cyclopentanol esters of N-trifluoroacetyl-alanine were partially resolved. Important control experiments were carried out by Gil-Av et al. to prove that indeed enantiomers were separated for the first time by enantio-GC [16]. The resolution was lost on an achiral stationary phase [29], and mixtures enriched in either the D- or L-amino acid enantiomer resulted in different peak areas reflecting the true enantiomeric ratio [16]. When the configuration of the CSP was reversed from L- to D-isoleucine, an expected peak reversal (i.e., an inversion of the elution order L vs. D of the enantiomers) was observed. These experiments unequivocally proved that enantiomers had been separated [16].

In a follow-up publication, 18 pairs of enantiomers of N-trifluoroacetyl-α-amino acid alkyl esters were resolved using the CSPs N-trifluoroacetyl-D-(or L-)isoleucine lauryl ester and N-trifluoroacetyl-L-phenylalanine cyclohexyl ester [31]. In all instances,

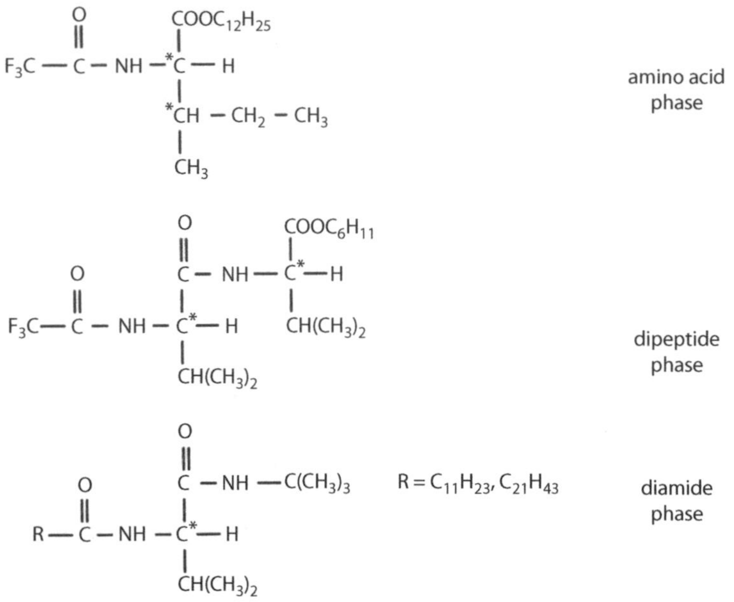

$$\underset{\substack{\displaystyle | \\ \overset{*}{C}H - CH_2 - CH_3 \\ | \\ CH_3}}{F_3C - \overset{O}{\overset{\|}{C}} - NH - \overset{COOC_{12}H_{25}}{\overset{|}{\underset{}{C}} - H}}$$

amino acid phase

$$F_3C - \overset{O}{\overset{\|}{C}} - NH - \underset{\substack{| \\ CH(CH_3)_2}}{\overset{*}{C}} - H \qquad \overset{O}{\overset{\|}{C}} - NH - \underset{\substack{| \\ CH(CH_3)_2}}{\overset{*}{C}} - H$$

dipeptide phase

$$R - \overset{O}{\overset{\|}{C}} - NH - \underset{\substack{| \\ CH(CH_3)_2}}{\overset{*}{C}} - H \qquad \overset{O}{\overset{\|}{C}} - NH - C(CH_3)_3 \qquad R = C_{11}H_{23}, C_{21}H_{43}$$

diamide phase

FIGURE 4.2   Structure of derivatized α-amino acid CSPs for enantio-GC.

it was found that the L-enantiomers of the derivatized α-amino acids eluted after the D-enantiomers on columns coated with CSPs having the L configuration. Gil-Av et al. noted that the difference in the free energies of solvation $-\Delta_{DL}\Delta G$ of the diastereomeric associates amounted to only 0.006 to 0.03 kcal/mole at the temperature of enantiosep-aration. The observed enantioselectivity was explained by hydrogen bonding between NH···F and NH···O=C– functions, whereby the latter contribution was considered to be more important. Whereas all previous enantioseparations were performed on an ana-lytical scale on coated glass capillary columns, another pioneering advance emerged shortly afterward using an enantioselective packed column for the semipreparative enantioseparation by GC [32], which paved the way for further developments [13]. Moreover, a chiroptical detector producing opposite optical rotatory dispersion (ORD) curves for the separated enantiomers was employed for the first time. In addition, a dipeptide derivative (Figure 4.2) was employed as a second-generation CSP. The results are shown in Figure 4.3, and the conditions are detailed in the caption.

The further development of the direct enantioseparation by GC by dipeptide or α-amino acid diamide phases has been described in detail [18,19,33]. As the C-terminal amino acid in dipeptide phases turned out to be unimportant for enantio-selectivity but provided only the necessary second amide bond for multiple hydro-gen bonding, it was substituted by a tert-butyl group to yield the versatile mono α-amino acid diamide selector N-lauroyl-L-valine-tert-butylamide of Feibush [34] (Figure 4.2). Later, the direct GC enantioseparation on optically active mesophase CSPs was achieved using mesomorphic chiral carbonyl-bis-α-amino acid esters [35].

In 1977, Frank, Nicholson, and Bayer linked the N-acyl-L-valine-tert-butylamide selector of Feibush [34] to a poly(dimethylsiloxane) backbone, thereby combining

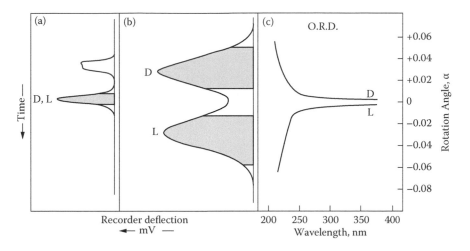

FIGURE 4.3   (a) Gas chromatogram of impure racemic D,L-*N*-TFA-alanine *tert*-butyl ester (shaded area; white area: impurity) on a 4 m × 6 mm (ID) packed column containing 20% achiral SE-30 on Chromosorb W at 125°C: no resolution occurs. (b) Gas chromatogram of the collected fraction (shaded area) corresponding to the second peak in (a) on a 2 m × 1 mm (ID) packed column containing 5% *N*-TFA-L-valyl-L-valine cyclohexyl ester (Figure 4.2) on Chromosorb W at 100°C: partial resolution occurs. (c) Optical rotatory dispersion (ORD) diagram of the two collected fractions corresponding to the shaded areas in (b) showing opposite rotation angles. (From E. Gil-Av and B. Feibush, *Tetrahedr. Lett.* 8 (1976) 3345–3347 with permission. © Elsevier.)

the inherent enantioselectivity of a CSP with the unique GC properties of silicones [36]. The *chiral* poly*silox*ane containing *val*ine diamide was termed *Chirasil-Val* (Figure 4.4, left). The novel polymeric CSP was commercialized as enantiomeric capillary columns coated with either Chirasil-L-Val or Chirasil-D-Val. The impressive simultaneous enantioseparation of all proteinogenic α-amino acids as $N(O,S)$-trifluoroacetyl-O-n-propyl esters on Chirasil-L-Val is depicted in Figure 4.5 [37]. Similar enantioseparations were also obtained with $N(O,S)$-pentafluoropropionyl-O-isopropyl esters of α-amino acids.

The pretreatment of borosilicate glass columns prior to coating with Chirasil-Val has been described in detail [38], and the immobilization property of Chirasil-Val on the glass surfaces was also studied [39]. In later work, Chirasil-Val-coated borosilicate glass capillary columns were substituted by Chirasil-Val-coated fused silica capillary columns, which were initially commercialized by Chrompack (Middelburg, The Netherlands) and are now available from Agilent (Santa Clara, CA, USA). A more straightforward approach to polymeric CSPs is based on the modification of cyanoalkyl-substituted polysiloxanes (XE-60, OV-225) [40,41].

Koppenhoefer et al. modified the chiral backbone in Chirasil-Val by variation of the loading and polarity of the CSP and by the introduction of rigid spacers [42–44]. In Chirasil-Val-C_{11}, a long undecamethylene spacer separates the valine diamide selector from the polymeric backbone [45]. In Chirasil-Val, the chiral moieties are statistically distributed about the polymer chain. A more ordered Chirasil-type CSP

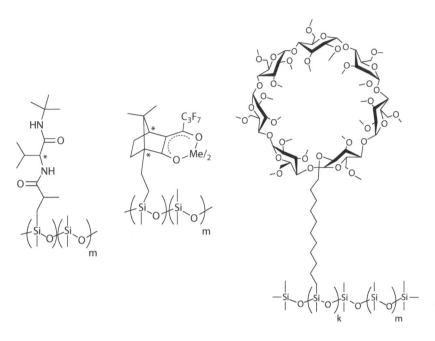

FIGURE 4.4 The poly(dimethylsiloxane)-anchored chiral selectors Chirasil-Val (left), Chirasil-metal (middle), and Chirasil-β-Dex-C₁₁ (right) for enantio-GC.

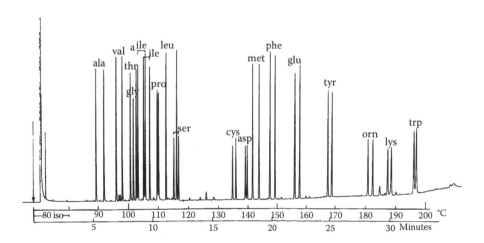

FIGURE 4.5 Simultaneous enantioseparation of proteinogenic *N(O,S)*-TFA α-amino acid *O-n*-propyl esters by GC on a 20 m × 250 μm ID glass capillary coated with Chirasil-L-Val by enantio-GC. The first eluted peak corresponds to the D-enantiomer. (From H. Frank, in: *Chirality and biological activity*, B. Holmstedt, H. Frank, and B. Testa, Eds., Liss, New York, 1990, Chapter 3, pp. 33–54, with permission.)

has been obtained by block condensation of 1,5-*bis*-(diethylamino)-hexamethyl-trisiloxane and 2',2',2'-trifluoroethyl-(3-dichloromethylsilyl)-2-methylpropionate followed by nucleophilic displacement of the functionalized polysiloxane with chiral amines and amino acids [46]. The immobilization of the CSPs by thermal [46] and radical-mediated cross-linking [47] has been studied, and the extent of radical-induced racemization was determined. s-triazine isopropyl esters of tripeptides such as L-valyl-L-valyl-L-valine were bonded to amino silicones and used as CSPs for the enantioseparation of a variety of racemates [48]. A diproline chiral selector was chemically linked to poly(dimethylsiloxane) and used as a CSP for the derivatization-free enantioseparation of several aromatic alcohols by enantio-GC [49].

Another type of Chirasil selectors has been obtained by substituting the *tert*-butyl amide group of Chirasil-Val by a cycloalkyl group and valine by tert-leucine [50]. A highly ordered supramolecular structure has been prepared by linking chiral L-valine-tert-butylamide moieties to the eight hydroxyl groups of a resorcine[4] arene supramolecular structure obtained from resorcinol and 1-undecanal. The calixarene-type selector was subsequently chemically linked via four spacer units to a poly(dimethylsiloxane) to give Chirasil-Calixval [51,52]. The synthesis of thiacalix[4]arenes with pendant chiral amines and their application as a CSP for the enantioseparation of derivatized amino acids, alcohols, and amines has also been described [53].

The enantioseparation of α-amino acids by hydrogen-bonding CSPs requires derivatization to increase the volatility or to introduce suitable functions for additional hydrogen bonding as well as for improving detection of trace amounts of enantiomers (e.g., by electron capture detection) [18,19]. The derivatization strategy should also assist the simultaneous enantioseparation of α-amino acids without extensive peak overlapping. At the outset of enantio-GC, Gil-Av et al. used a two-step derivatization strategy for α-amino acids [29,31], consisting of the formation of *N*-perfluoroacyl-*O*-alkyl esters, which proceeds without racemization at ambient temperature [54]. This two-step derivatization strategy has also frequently been used for achiral GC-MS analyses of α-amino acids [55]. For Chirasil-Val and related CSPs, *N*-trifluoroacetyl-*O*-methyl (or *O*-n-propyl and *O*-2-propyl) esters and *N*-pentafluoropropionyl-*O*-2-propyl esters are routinely employed [36,37,50] (cf. Figure 4.5). A very fast one-step derivatization procedure of the carboxylic group and all other reactive groups of α-amino acids has been developed by Hušek [56,57]. The use of alkyl chloroformates as derivatizing reagents leads to *N*(*O*)-alkoxycarbonyl alkyl esters of α-amino acids; the intermediate mixed anhydride is decarboxylated to the alkyl ester. The alkyl chloroformate approach bears a number of advantages: (a) The rapid one-step reaction can be carried out in aqueous solution without heating; (b) the cost of reagent is negligible; (c) the derivatized amino acids can easily be separated from the mixture using an organic solvent, thus reducing chemical contamination; and (d) the method can easily be automated [58]. A comprehensive study has been performed for the derivatization of α-amino acids suitable for stereochemical analysis on board spacecraft in space exploration missions of solar body environments [59]. The enantioseparation of α-amino acids derivatized with ethyl chloroformate has also been advanced in comprehensive two-dimensional (2D) GC [60].

4.3   CHIRAL STATIONARY PHASES BASED ON METAL COMPLEXES FOR COMPLEXATION ENANTIO-GC

In view of the high success of metal-catalyzed asymmetric synthesis, the complementary use of chiral metal coordination compounds as chiral selectors in enantio-GC commanded interest. Indeed, dicarbonylrhodium(I) 3-(trifluoroacetyl)-(1R)-camphorate (Figure 4.6), when dissolved in squalane and coated on a long stainless steel capillary column, displayed isotopic selectivity toward deuterated ethenes [61] and enantioselectivity toward the chiral olefin 3-methylcyclopentene [62]. Whereas the scope of enantioseparation of olefins was limited [63], the use of lanthanide(III) *tris*[3-(trifluoroacetyl)-(1R)-camphorates] [64], metal(II) *bis*[3-(trifluoroacetyl)-(1R)-camphorates], and metal(II) *bis*[3-(heptafluorobutanoyl)-(1R)-camphorates] (Figure 4.6) (metal = manganese, cobalt, nickel, and zinc) [65,66] as chiral selectors for the enantioseparation of nitrogen-, oxygen-, and sulfur-containing compounds emerged soon as a routine method in enantioselective complexation GC [5,67,68].

The classes of enantioseparated chiral compounds comprise cyclic ethers, esters, acetals, underivatized alcohols, and ketones, among them pheromones and essential oils [69,70]. Most of the compounds were not previously amenable to enantioseparation by hydrogen-bonding CSPs. The application of high-resolution glass or fused silica capillaries in complexation enantio-GC improved the state of the art, and squalane was later replaced by poly(dimethylsiloxane) as a useful solvent for the CSPs [71]. Significantly, the metal-containing CSPs were capable of enantioseparating some of the smallest chiral compounds, namely, alkyl-substituted aliphatic aziridines, oxiranes (Figure 4.7), and thiiranes [5,65]. The ability to enantioseparate small alkyl-substituted oxiranes found applications in the study of catalytic [72] and enzymatic epoxidations [73] as well as enzymatic kinetic resolutions of aliphatic epoxides [74]. Eleven 3-(heptafluorobutanoyl)-terpene ketones were probed

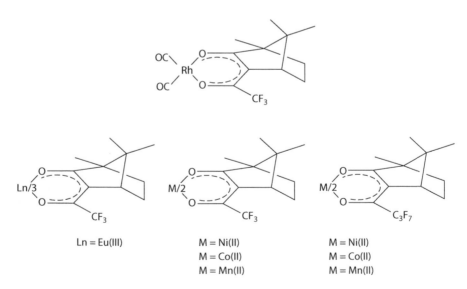

FIGURE 4.6   Structures of metal chelate CSPs for enantio-GC.

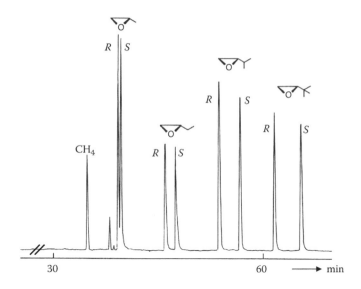

FIGURE 4.7 Enantioseparation of simple alkyloxiranes on a stainless steel capillary column (160 m × 0.4 mm ID) coated with 0.05 molal manganese(II) *bis*[3-(heptafluorobutanoyl)-(1*R*)-camphorate] in squalane at 60°C by complexation enantio-GC. (From V. Schurig, and R. Weber, *J. Chromatogr.* 217 (1981) 51–70 with permission. © Elsevier.)

as optically active *bis*-chelating ligands of nickel(II) in complexation enantio-GC, among them CSPs based on 3- and 4-pinanone, 4-methyl-3-thujone, carvone, pulegone, menthone, and isomenthone [75].

The low-temperature stability of the CSPs was later improved by chemically linking the nickel(II) chelate to poly(dimethylsiloxane) affording polymeric Chirasil-nickel (Figure 4.4, middle, Me = Ni) [76] in resemblance to the CSP Chirasil-Val [36]. Novel poly(dimethylsiloxane)-anchored Chirasil-nickel(II) CSPs, that is, 3-(perfluoroalkanoyl)-(1*R*)-camphorate nickel(II)-*bis*[(1*R*,4*S*)-3-trifluoromethanoyl-10-propylenoxycamphor]-polysiloxane and nickel(II)-*bis*[(1*R*,4*S*)-3-heptafluorobutanoyl-10-propylenoxycamphor]-polysiloxane, were synthesized, characterized, and immobilized (Figure 4.8) [77,78]

With the advent of modified CDs (Section 4.4), the practical use of complexation enantio-GC has vanished. The method, however, offered important insights into the mechanisms of chirality recognition in the realm of metal organic chemistry. Important information is gained from five peak parameters: (a) peak retention as a thermodynamic measure of the coordination equilibrium (chemoselectivity K, $-\Delta G$); (b) peak separation as a thermodynamic measure of chirality recognition (enantioselectivity $-\Delta_{D,L}\Delta G$); (c) peak assignment as a correlation of retention and chirality (assignment of molecular configuration D vs. L); (d) peak ratio as a measure of the enantiomeric composition (enantiomeric ratio er); and (e) peak coalescence as a kinetic measure of interconversion of enantiomers during separation (enantiomerization rate constant k_{inv}, $\Delta G^{\#}$) [65]. Various coalescence phenomena [79], including the dynamic process of enantiomerization [65,80], as well as four different enantioselective processes [72,81] were discovered in complexation

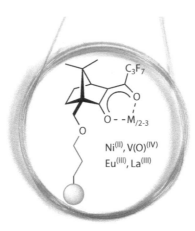

FIGURE 4.8 Chirasil-metal CSP derived from 3-heptafluorobutanoyl-(1*R*)-camphorate and coated on a fused-silica capillary employed in complexation enantio-GC. (From M. J. Spallek, G. Storch, and O. Trapp, *Eur. J. Org. Chem.* 21 (2012) 3929–3945 with permission.)

enantio-GC. In complexation enantio-GC, the first interconversion profiles of enantiolabile compounds on chiral metal complexes were observed (Figure 4.9, bottom) [65,80]. The terms *dynamic* and *enantiomerization* were introduced into enantio-GC, and a simulation algorithm based on the theoretical plate model was developed [80,82,83]. Thereby, the *principle of microscopic reversibility* requires that the rate constants of the forward interconversion (k^s_1) and that of the backward interconversion (k^s_{-1}) in the stationary phase differ in the presence of the chiral selector when enantiomeric separation occurs ($K_B > K_A$) (Figure 4.9, top). The comparison of an experimental and simulated interconversion profile for homofuran on Chirasil-nickel (Figure 4.4, middle, Me = Ni) is depicted in Figure 4.10 [82]. The research on interconverting enantiomers has greatly been expanded subsequently [5] and has extensively been reviewed also for other chromatographic methods [84–86].

In complexation enantio-GC, enantiomers have been separated in less than 20 s on Chirasil-nickel (Figures 4.4 and 4.11) [76]. Very fast enantioseparations in the second column are the prerequisite for applications in comprehensive two-dimensional gas chromatography (GCxGC).

In the realm of emerging supramolecular chromatography [87], a novel chiral three-dimensional (3D) metal open-framework crystalline material obtained from (1*R*)-camphoric acid, 1,4-benzenedicarboxylate, and 4,4′-trimethylenedipyridine has been dynamically coated on a fused silica capillary (2 m × 0.75 mm ID) [88]. According to Xie et al., the CSP possesses three elements of enantiopure features: (a) a 3D intrinsically homochiral net, (b) induced homohelicity, and (c) molecular chirality associated with (1*R*)-camphoric acid. The enantiomers of the terpenoid compounds citronellal and limonene and the alcohols 1-phenylethanol, 1-phenyl-1,2-ethanediol, and 2-amino-1-butanol (as trifluoroacetyl derivatives) were nearly baseline separated by enantio-GC [88]. The racemic α-amino acids glutamic acid and proline were resolved as trifluoroacetyl isopropyl esters on the metal organic framework (MOF) with α = 1.42 as compared to Chirasil-Val of α = 1.046 and 1.003, respectively [88].

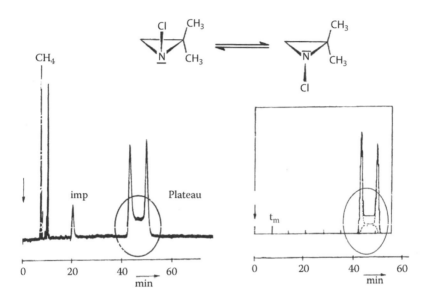

FIGURE 4.9   Top: Principle of microscopic reversibility as applied for enantiomerization in enantio-GC [80]. Bottom: Plateau formation due to enantiomerization (distorted elution profile caused by molecular inversion at the nitrogen atom) of 1-chloro-2,2-dimethylaziridine on complexation GC on nickel(II) *bis*[3-(trifluoroacetyl)-(*1R*)-camphorate] in squalane at 60°C. Left: Experimental trace; right: simulated trace. (From W. Bürkle, H. Karfunkel, and V. Schurig, *J. Chromatogr.* 288 (1984) 1–14 with permission. © Elsevier.)

In case the theoretical plate count can be improved for this enantioselective gas-solid chromatographic system, very good prospects for chiral MOFs, as important corollary to the traditional metal chelate CSPs (Figure 4.6), can be envisioned in complexation enantio-GC.

4.4   CHIRAL STATIONARY PHASES BASED ON DERIVATIZED CYCLODEXTRINS FOR INTER ALIA INCLUSION ENANTIO-GC

Modified cyclodextrins (CDs) have become the most important and widely used CSPs for the analytical separation of volatile chiral compounds by enantio-GC [6,89–94]. The early work of Sybilska et al. [95–97], employing underivatized

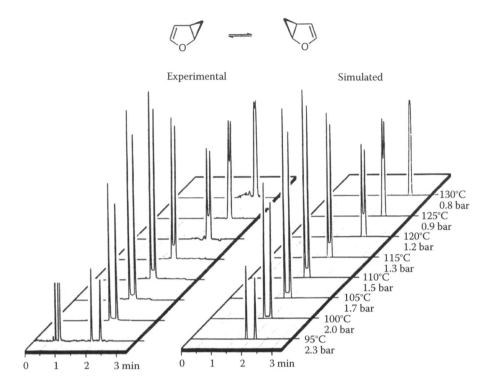

FIGURE 4.10 Comparison of experimental (left) and simulated (right) interconversion profiles due to enantiomerization of homofuran at elevated temperature by dynamic enantio-GC on a 10 m × 0.1 mm (ID) fused silica column coated with 0.25 μm Chirasil-nickel. (From V. Schurig et al., *Chem. Ber.-Recueil* 125 (1992) 1301–1303 with permission.)

α-cyclodextrin hydrate in formamide for the enantioseparation of terpenoic hydrocarbons (α- and β-pinene, *cis*- and *trans*-pinane and carene), started an impressive development of enantio-GC employing derivatized CDs despite the rather low efficiency inherent in the use of a packed column [98]. CDs represent a homologous series of cyclic oligosaccharides in which six, seven, or eight α-D-glucopyranose units are connected via α-(1 → 4)-glycosidic bonds. CDs form cavities with a narrow and wide entrance. The different reactivity of the 2-, 3-, and 6-hydroxy groups has been exploited since 1987 by regioselective alkylation and acylation [89], thus giving rise to a host of differently derivatized CDs [27,90,91,99]. The only disadvantage of CDs resides in their exclusive availability in the D-configuration of the glucose building blocks. It should be realized that the absence of the L-form prevents the possibility of reversing the sense of enantioselectivity by employing the CSP with opposite configuration, which usually represents an important tool for validation purposes.

Enantioselective gas-liquid chromatography requires a fluidic CSP that can be obtained by either dissolving solid CD derivatives in a suitable polysiloxane or by employing CD derivatives that are liquid at ambient temperature. The former approach was introduced by Schurig and Nowotny, who adopted strategies of complexation

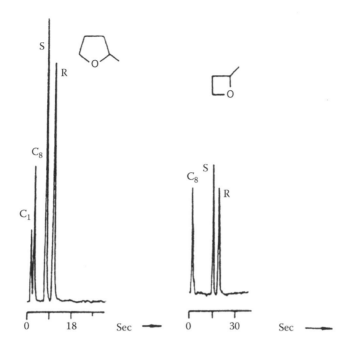

FIGURE 4.11   Fast enantioseparation of cyclic ethers on a miniaturized 1.5 m × 0.05 mm (ID) fused silica column coated with Chirasil-nickel (Figure 4.4, middle, Me = Ni) by complexation enantio-GC. Left: 2-Methyl-tetrahydrofuran, 115°C, 2.0 bar dinitrogen. Right: 2-Methyl-oxetane, 140°C, 2 bar dinitrogen. (From V. Schurig, D. Schmalzing, and M. Schleimer, *Angew. Chem. Int. Ed. Engl.* 30 (1991) 987–989 with permission.)

enantio-GC [71]. They consequently dissolved peralkylated CDs in semipolar polysiloxanes (e.g., OV-1701), thus combining enantioselectivity with the ideal coating properties of silicones in GC [100,101]. With CDs in the undiluted form, König et al. [102] used liquid per-*O*-pentylated and 3-*O*-acylated-2,6-di-*O*-pentylated α-, β-, and γ-CDs [91], whereas Armstrong et al. employed liquid permethylated 2-hydroxypropyl and pentylated/acylated CDs [103,104]. Two earlier reports described the use of undiluted permethylated β-CD in a supercooled state below its melting point for enantioseparations, albeit column efficiency was unsatisfactory [105–107].

The strategy of Schurig and Nowotny to dilute modified CDs in semipolar polysiloxanes [100] emerged soon as the most frequently used methodology in enantio-GC [11], and it was subsequently adopted also by König et al. for *n*-pentylated CDs [20]. The use of diluted CSPs bears a number of advantages [108–110]: (a) high melting points of CDs and phase transitions of CDs play no role; (b) multicomponent (mixed) CD-based CSPs can be employed; and (c) the universal coating properties of silicones for producing high-resolution and high-efficiency capillary columns by enantio-HRC-GC are maintained. A test mixture has been proposed for permethylated β-cyclodextrin in enantio-GC that covers the whole polarity spectrum (Figure 4.12) [108]. Only the highly polar free carboxylic acid shows some expected peak distortion. In Scheme 4.1, modified CDs frequently used as CSPs in enantio-GC are listed.

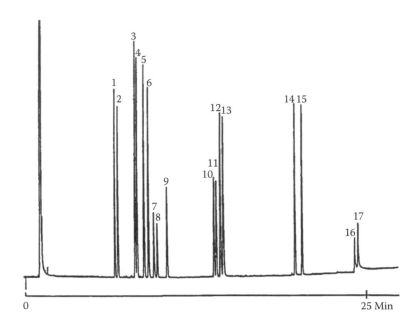

1	α–Pinene	10	γ–Valerolacton
2	α–Pinene	11	γ–Valerolacton
3	1R (+) –trans Pinane	12	1–Phenylethylamine
4	1S (–) –trans Pinane	13	1–Phenylethylamine
5	1S (–) –cis Pinane	14	1–Phenylethanol
6	1R (+) –cis Pinane	15	1–Phenylethanol
7	2, 3–Butanediol	16	2–Ethylhexanoic acid
8	2, 3–Butanediol	17	2–Ethylhexanoic acid
9	2, 3–Butanediol (meso)		

FIGURE 4.12 Simultaneous enantioseparation of apolar-to-polar chiral compounds ("Schurig test mixture" [108]) on permethylated β-cyclodextrin diluted in OV-1701 on a 25 m × 0.25 mm (ID) × 0.25 μm (film thickness) fused-silica capillary by enantio-GC (70°C for 5 min followed by 3°C/min, 0.65 bar dihydrogen [overpressure]). (Courtesy Sabine Mayer.)

Between 1988 and 2000, peralkylated β-CDs represented the most popular CSPs applied in enantio-GC (488 applications) [22]. Jaus and Oehme found a strong dependence of the purity status of permethylated β-CD on the enantioselectivity for chiral organochlorines [111]. In undermethylated specimens, CD desymmetrization and hydrogen-bonding effects can have an impact on enantioselectivity. Consequently, Cousin et al. synthesized eicosa-*O*-methyl-β-cyclodextrins that were devoid of one methyl group in the hydroxyl positions 2, 3, and 6 of β-CDs [112]. For various chiral 5-alkyl-5-methyl hydantoins, an influence of the free hydroxyl groups at the 2- and 3-positions at the wide side of the cavity on enantioselectivity, but not at the 6-position at the narrow side of the cavity, was observed vis-à-vis 2,3,6-heneicosa-*O*-methyl-β-cyclodextrin [112].

A comprehensive collection of practical enantioseparation factors α of different classes of chiral compounds measured on octakis(6-*O*-methyl-2,3-di-*O*-pentyl)

Heptakis(2,6-di-O-methyl-3-O-pentyl)-β-cyclodextrin
Octakis(2,6-di-O-methyl-3-O-pentyl)-γ-cyclodextrin
Octakis(6-O-methyl-2,3-di-O-pentyl)-γ-cyclodextrin (Lipodex G)
Hexakis(2,3,6-tri-O-pentyl)-α-cyclodextrin
Octakis(2,3,6-tri-O-pentyl)-γ-cyclodextrin
Hexakis(3-O-acetyl-2,6-di-O-pentyl)-α-cyclodextrin
Heptakis(2,3,6-tri-O-pentyl)-β-cyclodextrin
Heptakis(3-O-acetyl-2,6-di-O-pentyl)-β-cyclodextrin
Octakis(3-O-butanoyl-2,6-di-O-pentyl)-γ-cyclodextrin (Lipodex E)
Hexakis(per-O-2-hydroxypropyl)-per-O-methyl)-α-cyclodextrin (PMHP-α-CD)
Heptakis(per-O-2-hydroxypropyl)-per-O-methyl)-β-cyclodextrin (PMHP-β-CD)
Hexakis(2,6-di-O-pentyl)-α-cyclodextrin (dipentyl-α-CD)
Heptakis(2,6-di-O-pentyl)-β-cyclodextrin (dipentyl-β-CD)
Heptakis(2,6-di-O-pentyl-3-O-trifluoroacetyl)-β-cyclodextrin (DPTFA-β-CD)
Octakis(2,6-di-O-pentyl-3-O-trifluoroacetyl)-γ-cyclodextrin (DPTFA-γ-CD)
Heptakis(2,3-di-O-acetyl-6-O-*tert*-butyldimethylsilyl)-β-cyclodextrin
Heptakis(2,3-di-O-methyl-6-O-*tert*-butyldimethylsilyl)-β-cyclodextrin

SCHEME 4.1 Derivatized cyclodextrins for enantio-GC complementing the use of per-methylated β-cyclodextrin and Chirasil-β-Dex.

-γ-cyclodextrin (Lipodex G) [113], on heptakis(2,6-di-O-methyl-3-O-pentyl)-β-cyclodextrin and octakis(2,6-di-O-methyl-3-O-pentyl)-γ-cyclodextrin [114], and on octakis(3-O-butanoyl-2,6-di-O-pentyl)-γ-cyclodextrin (Lipodex E) [115] was compiled by König. The influence of the substituents in positions 2, 3, and 6 of the glucose building blocks in the CDs has been discussed by Mosandl et al. [116]. Bulky substituents in the 6-position of CDs are beneficial for chirality recognition [117,118]. Per-*tert*-butyldimethylsilyl (TBDMS)-β-cyclodextrin diluted in the polysiloxane PS-086 was initially used as a versatile CSP for the GC enantioseparation up to 250°C [119]. Since not all hydroxyl groups are silylated in different commercial batches, reproduction of enantioseparation is difficult and may lead occasionally even to peak reversals [11]. Three related CD derivatives carrying the TBDMS groups only in the 6-position on the narrow site of the CD cavity have emerged as important CSPs in enantio-GC: heptakis(2,3-di-O-methyl-6-O-*tert*-butyldimethylsilyl)-β-cyclodextrin [120], heptakis(2,3-di-O-ethyl-6-O-*tert*-butyldimethylsilyl)-β-cyclodextrin [121], and heptakis(2,3-di-O-acetyl-6-O-*tert*-butyldimethylsilyl)-β-cyclodextrin [122]. A comprehensive collection of enantioseparation factors α of different classes of chiral compounds measured on heptakis(2,3-di-O-methyl-6-O-*tert*-butyl dimethylsilyl)-β-cyclodextrin [120] was compiled by Mosandl et al. [123,124]. To evaluate the role of the substitution pattern in positions 2 and 3 on the enantioselectivity, the asymmetrically substituted CSPs heptakis(3-O-ethyl-2-O-methyl-6-O-*tert*-butyldimethylsilyl)-β-cyclodextrin (MeEt-CD) and heptakis(2-O-ethyl-3-O-methyl-6-O-*tert*-butyldimethylsilyl)-β-cyclodextrin (EtMe-CD) where compared with the symmetrical substituted CSPs heptakis(2,3-di-O-methyl-6-O-*tert*-butyldimethylsilyl)-β-cyclodextrin (MeMe-CD) and heptakis(2,3-di-O-ethyl-6-O-*tert*-butyldimethylsilyl)-β-cyclodextrin (EtEt-CD) [125]. It was found that the asymmetrically substituted methyl/ethyl CDs, compared to the symmetrical substituted dimethyl or diethyl CDs, exhibited better enantioselectivity for an increased number of enantioseparated chiral compounds, and it showed that the substitution pattern in positions 2 and 3 of the CD governs enantioseparation [125]. Likewise, octakis(3-*O*-acetyl-2-*O*-methyl-6-*O*-*tert*-butyldimethylsilyl)-γ-cyclodextrin and octakis(2-*O*-acetyl-3-*O*-methyl-6-*O*-*tert*-butyldimethylsilyl)-γ-cyclodextrin were compared with octakis(2,3-di-*O*-methyl-6-*O*-*tert*-butyldimethylsilyl)-γ-cyclodextrin and octakis(2,3-di-*O*-acetyl-6-*O*-*tert*-butyldimethylsilyl)-γ-cyclodextrin. The asymmetrically substituted octakis(3-*O*-acetyl-2-*O*-methyl-6-*O*-*tert*-butyldimethylsilyl)-γ-cyclodextrin showed the highest enantioselectivity among the four CDs investigated [126]. The exchange of just one single methyl group in position 3 of heptakis(2,3-di-*O*-methyl-6-*O*-*tert*-butyldimethylsilyl)-β-cyclodextrin for an acetyl group increased the enantioseparation factor α of α-hexachlorocyclohexane from 1.14 to 1.53 [127]. When compared to permethyl-β-CD, the TBDMS derivatives of β-CD showed a higher solubility in apolar polysiloxanes and are preferred diluted CSPs (e.g., in OV-1701) in enantio-HRC-GC.

In diluted systems, the enantioseparation factor α is rendered concentration dependent due to the two different retention mechanisms arising (a) from the presence of the achiral polysiloxane and (b) from the CD diluted into it, both comprising the total CSP [128–130]. A theoretical treatment has shown that α does not linearly increase with the CD concentration but reaches an optimum often at a low CD concentration [129]. Hence, no gain in enantioselectivity above an optimum value is obtained. Thus, the exclusive use of *undiluted* CDs [91,130] is therefore increasingly discontinued.

An obvious extension of the dilution approach consists of the fixation of the CD to a poly(dimethylsiloxane) backbone by a permanent chemical linkage yielding a CSP reminiscent of Chirasil-Val [36]. The synthesis of covalently linked permethylated β-CD (Chirasil-β-Dex) (Figure 4.4, right) has been realized independently by two groups [76,131,132]. The application of these polymeric phases in HRC-GC offers such advantages as (a) the use of an apolar poly(dimethylsiloxane) as matrix for CDs; (b) a high degree of inertness, allowing the fast analysis of polar racemates; (c) the use of low to high CD concentrations; (d) the immobilization of the CSP by thermal treatment, leading to solvent tolerance (in online injection or by rinsing of columns); (e) the compatibility with a very wide temperature range (–25°C to 250°C); and (f) the strong reduction of column bleeding necessary for interfacing with mass spectrometry (MS) (Section 4.6).

The thermal immobilization of Chirasil-Dex (and of similar CSPs) on all kinds of silica surfaces (glass, fused silica, silica particles) offers universal applicability of this CSP in different chromatographic and electromigration modes, including an enantioselective unified approach in which a single miniaturized capillary column (1 m × 0.05 mm ID) coated with Chirasil-β-Dex has been used for enantioseparation of a single racemate (hexobarbital) by GC, supercritical fluid chromatography, liquid chromatography, and capillary electrochromatography [133]. In the synthesis of Chirasil-Dex, a monoalkenyl (e.g., allyl, 1-pentenyl, 1-octenyl) residue is introduced into one hydroxyl group of the CD followed by permethylation and linking the selector to poly(hydridodimethylsiloxane) via hydrosilylation with a platinum catalyst [76,131]. The statistical synthesis involving chromatographic purification furnishes preferentially the monosubstituted 2-alkenylated product [134] and not, as indicated earlier in the literature, the 6-alkenylated regioisomer [76,131,132]. Via hydroxyl group protection chemistry, all three regioisomeric Chirasil-β-Dex CSPs (with 2-, 3-, and 6-octamethylene spacers) have been obtained and compared [134]. They exhibit nearly identical GC enantioselectivity for the investigated racemates. An improved CSP is Chirasil-β-Dex-C_{11}, which contains a long undecamethylene spacer between CD and poly(dimethylsiloxane) [45]. The CSP Chirasil-γ-Dex refers to poly(dimethylsiloxane)-linked Lipodex E [135]. Another immobilization strategy to link β-CD to siloxanes has been advanced by Armstrong et al. [136] and by Bradshaw et al. [137,138].

The option to mix different CD selectors in one CSP has been proposed for enantioseparations by GC [90]. Subsequently, in a number of publications, the use of mixed binary CD selector systems for the enantioseparation of different classes of chiral compounds has been described [139–143]. The mixture of heptakis (2,6-di-O-methyl-3-O-pentyl)-β-cyclodextrin and octakis(2,6-di-O-methyl-3-O-trifluoracetyl)-γ-cyclodextrin dissolved in the polysiloxane OV-1701 (ratio 2:1:7) turned out to be useful for the enantioseparation of chiral terpenes and lactones [140]. Recently, a comprehensive enantioselectivity spectrum toward racemic unfunctionalized alkanes and racemic derivatized α-amino acids as well as to other classes of racemic compounds such as lactones, diols, secondary alcohols, ketones, and terpenes has been realized by diluting the two versatile CD selectors Lipodex G [113] and heptakis(2,3-di-O-methyl-6-O-$tert$-butyldimethylsilyl)-β-cyclodextrin [120] in the polysiloxane PS-086 [144].

The GC enantioseparation of a wide variety of racemic compounds of different classes of compounds on modified CDs usually displays rather small enantioseparation factors ($1.02 < \alpha < 1.20$) (Figure 4.12), corresponding to a low enantioselectivity of $-\Delta_{D,L}(\Delta G)$ in the range of 0.014–0.140 kcal/mol at 100°C. The low enantioselectivity is beneficial for fast enantiomeric analysis. But, low Gibbs energy differences render any molecular modeling studies dubious. Mixed retention mechanisms in enantio-GC involving CDs have been identified by Armstrong et al. [145]. The inclusion mechanism is not always essential for chirality recognition since enantioseparations have also been observed in enantio-GC with modified linear dextrins ("acyclodextrins") devoid of a molecular cavity (Figure 4.13) [146,147].

Stable room temperature ionic liquids (RTILs) solubilize polar molecules such as CDs, and their wetting ability and viscosity allow them to be coated onto fused silica capillaries [148]. Ionic liquid stationary phases have been associated with a dual nature [148]: (a) They may act as a low-polarity stationary phase to nonpolar compounds, and (b) they retain polar molecules with strong proton donor groups whereby the nature of the anion can have a significant effect on both the solubilizing ability and the selectivity of ionic liquid stationary phases. Recently, it has been found that RTILs can be used as stable stationary phases entailed with unusual selectivity. Two approaches on the use of ionic liquids in enantio-GC have been realized [149,150]: (a) A chiral selector can be dissolved in an achiral ionic liquid [151], or (b) the ionic liquid itself can be chiral [149]. Compounds that have been separated using an ionic liquid chiral selector include alcohols, diols, sulfoxides, epoxides,

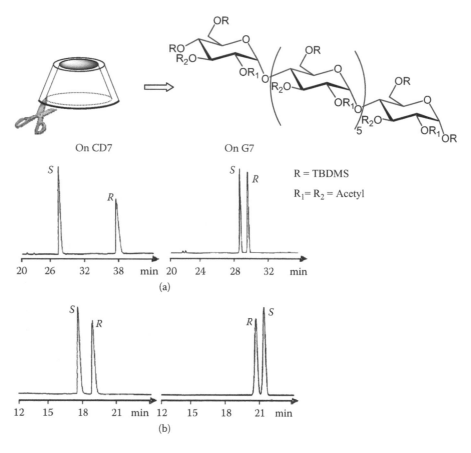

FIGURE 4.13 Comparison of the enantioseparation of the racemic pentafluoropropanoyl derivatives of (a) 1-(phenyl)ethylamine and (b) 1-(cyclohexyl)ethylamine on heptakis(2,3-di-O-acetyl-6-O-*tert*-butyldimethylsilyl)-β-cyclodextrin [122] versus heptakis[(2,3-di-O,4''-O)-acetyl-(1'-O,6-O)-*tert*-butyldimethylsilyl]-maltoheptaose (G7), on a 25 m × 250 μm ID fused-silica capillary coated with 0.25 μm of each CSP at 100°C and 100 kPa dihydrogen by enantio-GC. (From G. Sicoli et al., *Chirality* 19 (2007) 391–400 with permission.)

and acetylated amines. Because of the synthetic nature of these chiral selectors, the configuration of the stereogenic center can be controlled and altered for mechanistic studies and reversing enantiomeric retention. Alternatively, permethylated β-CD has been dissolved in the achiral ionic liquid [N-butyl-N-methylimidazolium]Cl⁻, and the mixture was coated on a capillary column [151]. The observed enantioselectivities and efficiencies differed from that observed with permethylated β-CD dissolved in polysiloxane. An enhancement of separation efficiency was accompanied by a loss of enantioselectivity, and it was explained by the imidazolium cation occupying the CD cavity, preventing inclusion complexation, which is crucial for chirality recognition [151]. This work was later extended to the use of ionic CDs, that is, permethylated mono-6-(butylimidazolium)-β-cyclodextrin and permethylated mono-6-(tripropylphosphonium)-β-cyclodextrin paired with iodide and triflate anions and dissolved in di- and tricationic ionic liquids, with improved column performance and good enantioselectivities [152]. The polar racemate 1-phenyl-2,2,2-trifluoroethanol showed longer retention and an improved peak shape on the ionic CSP compared to permethylated β-CD in polysiloxane [152].

Permethylated α-, β-, and γ-CDs have also been incorporated in a sol-gel matrix and used for the enantioseparation of terpenoids [153], and permethylated β-CD was also applied in a poly(oxyethylene) matrix [154]. The use of permethylated cyclofructans as novel CSPs in enantio-GC has been advanced by Armstrong et al. [155].

Commercial vendors of fused silica capillary columns coated with modified CDs are listed in He and Beesley [2], Vetter and Bester [11], and in a book by Schreier et al. [27].

4.5 THE USE OF MIXED CSPS CONSISTING OF α-AMINO ACID DIAMIDES AND MODIFIED CYCLODEXTRIN SELECTORS IN ENANTIO-GC

At present, there is no universal CSP known that would cover the whole range of chiral classes of compounds amenable to contemporary enantio-GC. In a search for homochirality in space research [156,157], chiral hydrocarbons and derivatized amino acids should preferentially be determined on a single capillary column [158]. Although dual chiral selector systems can be realized by combining two columns with different selectors in a tandem-column arrangement [158,159], the mixing of two selectors in one CSP in a single-column format [158] would meet the aforementioned requirement. When two enantioselective selectors are employed, their individual contributions to chirality recognition may lead to enhancement ("matched case") or to compensation ("mismatched case") of enantioselectivity [160]. In the absence of cooperative effects, the enantioselectivity obtained on a mixed binary chiral selector system is smaller than that of a single chiral selector system containing the more enantioselective selector. Therefore, it may be unfavorable, as inferred by Pirkle and Welch [159], to combine different selectors in one CSP. Yet, for practical purposes, the combination of chiral selectors with complementary enantioselectivity toward enantiomers of very different classes of racemic compounds in one CSP may result in a broader spectrum of enantioselectivity compared to those provided by either of the single-selector CSPs [3].

Three different approaches of mixed binary selector systems in enantio-GC have been described: (a) The two CSPs Chirasil-Calixval [51,52] and Chirasil-β-Dex [131] were bonded to a poly(dimethylsiloxane) matrix to furnish Chirasil-Calixval-Dex [161,162]; (b) the two CSPs Chirasil-Val-C_{11} and Chirasil-β-Dex C_{11} were bonded to poly(dimethylsiloxane) to furnish Chirasil-DexVal-C_{11} [45]; and (c) the CSP Chirasil-Val-C_{11} [45] was doped with Lipodex E [115] to furnish Chirasil-Val(γ-Dex) [163]. This mixed phase in which one CSP is dissolved in another CSP was found to have improved enantioselectivity toward proline and aspartic acid (as N-trifluoroacetyl ethyl or methyl esters) in comparison to Chirasil-Val [36]. Furthermore, the presence of Lipodex E [115] extended the scope of enantioseparations achievable on Chirasil-Val toward underivatized alcohols, terpenes, and many other racemic compounds.

For space explorations, one single capillary column should be capable to enantiosep-arate chiral saturated hydrocarbons and derivatized α-amino acids. An advance to the challenge is provided by the mixed CSP Chirasil-DexVal-C$_{11}$, which can simul-taneously enantioseparate the saturated hydrocarbons *trans*-1,2- and *trans*-1,3-dimethylcyclohexane, 3,5-dimethylcyclohex-2-en-1-one, 1-(2-methylphenyl)ethanol, and α-amino acids as *N*-trifluoroacetyl ethyl esters (Figure 4.14) [45].

A viable strategy to combine the enantioselectivities of hydrogen-bonding and inclusion-type selectors in a single CSP consists of linking L-valine moieties *directly* to the permethylated β-CD selector in similarity to Chirasil-Calixval [49,50]. Whereas the selector Valdex (heptakis[6-*O*-(*N*-acetylyl-L-valine-*tert*-butylamide)-2,3-*O*-methyl]-β-cyclodextrin) represented a versatile chiral solvating agent (CSA) for the nuclear magnetic resonance (NMR) spectroscopic discrimination of enantiomers [164], Chirasil-Valdex showed no remarkable enantioseparations of different classes of chiral compounds in enantio-GC. However, promising results have been obtained on a selector that carries only a single L-valine diamide moiety in the C6-position of per-methylated β-CD for enantioseparations by GC [165]. When the spacer length between the L-valine selector and β-CD was increased to avoid self-inclusion, whereby L-valine was bonded to the 6-position of the CD from either the *N*- or *C*-terminal end of the amino acid, an improvement of the enantioseparation factors α compared to the single-parent CSPs was observed for many derivatized α-amino acids [166].

4.6 HYPHENATED AND MULTIDIMENSIONAL APPROACHES IN ENANTIO-GC

Because of speed of analysis, high efficiency, high resolution, and absence of liquid mobile phases, enantio-GC is ideally suited for various coupling techniques and multidimensional approaches. These ancillary techniques are sometimes aided by the ease of miniaturization of enantio-GC [167,168].

When both high sensitivity and analyte identification are required, the coupling of gas chromatography and mass spectrometry (enantio-GC-MS) is the method of choice [3,24]. A link exists between suitable derivatization of α-amino acids and their determination via enantio-GC-MS [58,59]. In the selected ion-monitoring mode (MS-SIM), single ions are selected, thereby prolonging the detection time of these ions and thus increasing the signal-to-noise ratio. The four stereoisomers of *E,Z*-2-ethyl-1,6-dioxaspiro[4.4]nonane, chalcogran (the aggregation pheromone of the bark beetle *Pityogenes chalcographus*), were separated by complexation enantio-GC on nickel(II) *bis*[3-(heptafluorobutanoyl)-(*1R*)-camphorate] and detected at the parts-per-million level [68]. Unnatural D-amino acids were determined as *N(O*-pentafluoropropanoyl-*O*-2-propyl esters in mammals on Chirasil-Val in the enantio-GC-MS(SIM) mode [169]. Enantio-GC-MS(SIM) utilizing modified CDs was also used (a) for the chiral analysis of γ-lactones in milk products [170], (b) for the detection of an enantiomeric bias of the atropisomeric polychlorinated biphenyl PCB (polychlorinated biphenyl) 132 in human milk samples [171], (c) for the enantiomeric purity determination of *R*-limonene in citrus matrices [172], and (d) for the quantitative stereoisomeric composition of chi-ral hydrocarbon biomarkers 1-methyltetralin *cis*-1,2- and 1,3-dimethylindane in crude oil and coal samples at the 50 pg/g range [173]. A headspace enantio-GC-MS(SIM)

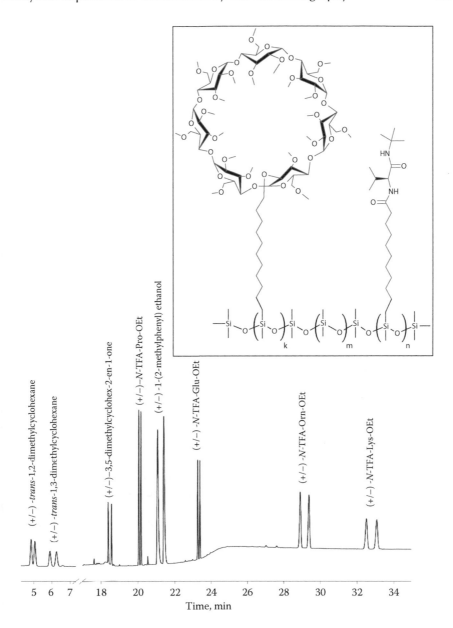

FIGURE 4.14 Simultaneous enantioseparation of apolar to polar chiral compounds on the mixed CSP Chirasil-Val-Dex-C_{11} [20% permethylated β-cyclodextrin and 15% valine-diamide on poly(dimethylsiloxane)] at 40°C/10 min isothermal, 10°C/min up to 170°C followed by 20 min isothermally by enantio-GC. (From P. A. Levkin et al., *Anal. Chem.* 78 (2006) 5143–5148 with permission. © American Chemical Society.)

study revealed that the racemic inhalational anesthetic isoflurane (2-chloro-2-(difluoromethoxy)-1,1,1-trifluoroethane) (Figure 4.15, top) showed a deviation of the enantiomeric ratio of the administered racemic composition from $S/R = 50/50$ to $S/R = 54/46$ during and after surgery in clinical patients [9,174].

The in situ deuteration and subsequent chiral analysis of α-amino acids by GC-MS(SIM) have been used as an indispensible tool to study the undesired racemization of α-amino acids occurring during the acid-catalyzed hydrolysis of peptides that can falsify the true enantiomeric composition of the building blocks of a peptide. When the hydrolysis is carried out in a fully deuterated medium (e.g., in 6 N D$_2$O/DCl), racemization during hydrolysis is accompanied by substitution of the hydrogen atom attached to the stereogenic carbon atom by deuterium, and the hydrogenated and deuterated species can be differentiated by MS. After hydrolysis, the amino acids are derivatized and analyzed by enantio-GC-MS(SIM) [175]. For each amino acid, only the characteristic ion containing the proton at the stereogenic carbon atom is considered, whereas the deuterated species arising from racemization are disregarded. The reliable determination of enantiomeric purities of L-α-amino acids in peptides up to 99.9% is thus possible [175]. This method has also been used to determine the rate constants of configurational inversion at the stereogenic carbon center of α-amino acids under harsh acid hydrolysis conditions (110°C, 6 N DCl) [175]. The time-dependent racemization of α-amino acids has been used for dating purposes of archeological artifacts containing α-amino acids [176]. An automated enantio-GC analysis system for α-amino acids has been described [177]. Freeze-dried body fluid, tissue, and food proteins were hydrolyzed, and the hydrolysates were automatically derivatized and analyzed on Chirasil-Val or Chirasil-γ-Dex [177]. Hyphenation of enantio-GC with ammonia chemical ionization MS has been described for the analysis of the nerve agent tabun [178]. Previously, the four stereoisomers of sarin have been separated by enantio-GC on Chirasil-L-Val [179].

A novel technique is concerned with enantiomeric analysis by the coupling of gas chromatography/^{1}H-NMR-spectroscopy (enantio-GC-^{1}H-NMR). Whereas the off-line NMR-spectroscopic investigation of stereoisomers separated by microscale-preparative capillary GC is straightforward [180], the first direct online coupling of enantio-GC-NMR has been demonstrated recently [8,181]. Thus, the chiral unfunctionalized saturated hydrocarbon 2,4-dimethylhexane (C*HMeEtiBu) was enantioseparated at 60°C on Lipodex G, and the 2D plot of the GC-^{1}H-NMR (400 MHz) experiment showed two separate clustered signals of the alkyl resonance adsorptions [181]. Previously, the diastereomeric *cis*- and *trans*-1,2-dimethylcyclohexanes were separated by GC on a 25 m × 250 μm ID fused silica capillary coated with 0.25 μm Chirasil-β-Dex and differentiated by ^{1}H-NMR (400 MHz) [182].

Enantio-GC is well suited for multiple-column operation. The presence of chiral compounds in multicomponent matrices will lead to the doubling of peaks when a suitable enantioselective column is used. The increased complexity of the elution pattern can be overcome by multidimensional approaches (i.e., by introducing chiral separation as an additional dimension to achiral GC). The conventional enantioselective technique involves the multidimensional heart-cut GC-GC approach (enantio-MDGC) first demonstrated by Schomburg et al. [183]. In this two-dimensional approach, the first nonenantioselective column coated with an achiral

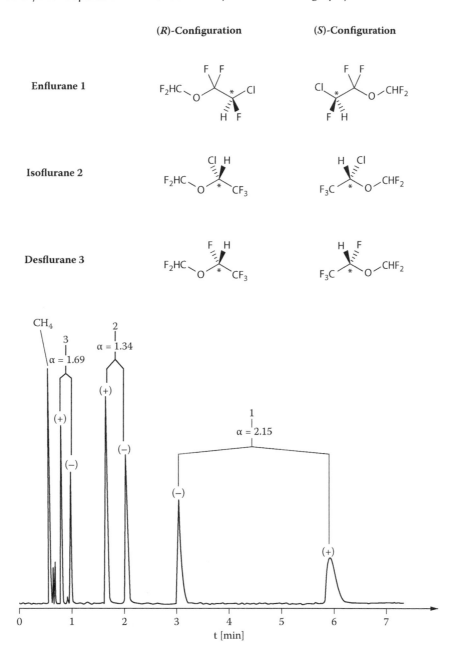

FIGURE 4.15 Enantioseparation of racemic inhalation anesthetics desflurane, isoflurane, and enflurane [27°C, 10 m × 0.25 mm ID fused-silica capillary coated with 0.18-μm immobilized Chirasil-γ-Dex/poly(dimethylsiloxane), 40%, w/w] by enantio-GC. (From H. Grosenick and V. Schurig, *J. Chromatogr.* 761 (1997) 181–193 with permission. © Elsevier.)

polar stationary phase is used to preseparate components of interest (first dimension), whereas in the second enantioselective column coated with a CSP, fractions of chiral analytes are enantioseparated after online transfer through a pneumatic or flow-controlled low-dead-volume heart–cut interface (second dimension). There are many applications in the field of terpenes and food chemistry [184–189].

The traditional one-column operation has been compared with enantio-MDGC for the enantioseparation of monoterpenes from essential oils [190]. A dual-column approach was used for the separation of all eight stereoisomers of the sex pheromone of a pine beetle, that is, derivatized dipropionol (2-hydroxy-3,7-dimethyl-heptadecane), containing three stereogenic centers. The achiral column preseparated diastereomers, whereas the enantioselective column based on polysiloxane-linked XE-60-L-valine-(S)-α-phenylethylamide [19] separated the enantiomers [191]. The analysis of soman stereoisomers involved thermodesorption/cold trap injection and 2D GC with an achiral precolumn followed by resolution of all four stereoisomers and an internal standard on a Chirasil-Val-coated capillary column [179,192]. From a complex mixture of atropisomeric PCBs present in the formulation Clophen A 60, one single peak was heart-cut in the first dimension and resolved into three peaks in the second dimension using an enantioselective fused-silica capillary coated with Chirasil-β-Dex. The peaks were assigned to PCB 151 (chiral and containing *two* chlorine substituents in ortho-positions: unresolved) and to PCB 132 (chiral and containing *three* chlorine substituents in ortho-positions: resolved) [193].

In the realm of flavor chemistry, enantioselective two-dimensional GC is frequently coupled with mass spectrometry (enantio-MDGC-MS) [194,195]. An example is provided by the stir-bar sorptive extraction (SBSE) and enantio-MDGC-MS-analysis of chiral flavor compounds in strawberries [196,197]. A refined enantioselective and multi-heart-cuts multidimensional GC-MS system has been used for the evaluation of chiral tea ingredients [198].

The development of comprehensive two-dimensional gas chromatography GCxGC provides an orthogonal separation of compounds on two columns using cryogenic modulation without the requirement of heart-cutting [199,200]. It requires fast mass-analyzing systems and devices for comprehensive data treatment. In principle, GCxGC works by transferring hundreds of contiguous heart-cuts from the first column (30 m × 0.25 mm ID, 0.25-μm coating) to the miniaturized second column (0.5–1 m × 0.1 mm ID, 0.1-μm coating), allowing generation of characteristic contour plots. If an achiral capillary column system is employed in the first dimension, a very fast enantioseparation is required in the second dimension (GC x enantio-GC) [201]. Only few enantioseparations are known today that fulfill this difficult requirement (Figure 4.11). The enantiomeric distribution of several monoterpene compounds in bergamot essential oil was reported as a demonstration of the method [201]. The column geometry can also be inverted [202–204]; that is, the first column represents an enantioselective separation system, whereas the second miniaturized achiral column is used to separate overlapping peaks and to obtain the two-dimensional contour plot (enantio-GCxGC). Heart-cut MDGC and GCxGC were compared for the analysis of perchlorinated biphenyls in foodstuff on a modified CD [205].

By enantioselective stopped-flow multidimensional gas chromatography (sf-MDGC), the configurational lability of a chiral compound can be evaluated at elevated temperatures with only a minute amount of the racemic sample [206,207]. Three columns are employed in series. The enantiomers are quantitatively separated online in the first enantioselective column. Either one of the pure enantiomers is then transferred into the second empty reactor column. By stopping the flow for the time t, enantiomerization (inversion of configuration) is allowed to proceed at the elevated temperature T in the gas phase. After the time t, the enantiomers are focused by cooling, and the flow is resumed. The enantiomeric ratio er is then determined in the third enantioselective column (or by backflushing into the first column). From T, t, and er, the rate constant of enantiomerization can easily be calculated. This method has been applied for the determination of activation barriers of molecular inversion of atropisomeric PCBs (chiral PCBs) [206,207], the chiral allene dimethyl-2,3-pentadienedioate [83,207], and the chiral nitrogen invertomers of 1-chloro-2,2-dimethylaziridine [207–209].

The online coupling liquid chromatography–gas chromatography (LC-GC) enables an effective sample fractionation in the LC step, even working in a reversed-phase (RP) mode, as well as an efficient and rapid GC separation. Considering the incompatibility between both analytical techniques (i.e., LC and GC), interfaces allowing the direct coupling are essential [210]. Enantioselective multidimensional LC-GC (enantio-MDLCGC) has been employed for the determination of enantiomeric compositions of essential oils [210–214].

The online coupling enantioselective multidimensional gas chromatography/isotope ratio mass spectrometry (enantio-MDGC-IRMS) has been used in the analysis of essential oils and in space research. The authenticity of genuine natural products is often evaluated by the determination of the enantiomeric excess (*ee*). However, the practice of fortifying (i.e., deliberately adding artificial synthetic enantiomers to essential oils) cannot be verified by this means. Mosandl therefore developed the enantio-MDGC-IRMS method based on the fact that natural and synthetic compounds show subtle differences in their isotopic ratios due to kinetic isotopic effects operating in various biosynthetic pathways [215]. The enantiomers eluted from the GC column are burned in a pyrolysis interface, and the $^{18}O/^{16}O$ or the $^{13}C/^{12}C$ ratio of the carbon dioxide formed is then determined by MS [188,216–220]. Also, biosynthetic routes or biogenetic pathways may be discerned by MDGC-IRMS [217,221]. A recent application was concerned with the genuineness assessment of mandarin and strawberry essential oils [222,223]. Enantioselective GC-IRMS has also been used to determine the isotopic ratio of carbon of extraterrestrial amino acids present in the Murchison meteorite [224].

4.7 DETERMINATION OF ABSOLUTE CONFIGURATION IN ENANTIO-GC

In stereochemistry, the knowledge of absolute molecular configurations is of paramount importance. Apart from a novel direct visualization of a chiral molecule [225], absolute configurations can be assigned by anomalous X-ray crystallography [226] and by chiroptical methods [227]. These methods rely on theoretical assumptions or quantum mechanical calculations [225].

By enantioselective NMR spectroscopy and chromatography, only relative configurations can be determined. In enantio-GC, molecular configurations can be determined by *direct evidence* through the simultaneous injection of a reference compound with an established stereochemistry. By *indirect evidence*, it is tempting to correlate the configuration of the enantiomers of structurally related compounds (congeners, homologues) with their order of elution from a given CSP. Indeed, on L-valine diamide selectors, the derivatized L-enantiomers of all proteinogenic α-amino acids were found to elute as the second peak [7]. By coincidence, the same elution order was also observed for Chirasil-γ-Dex (Lipodex E, comprised of D-glucose moieties, anchored to polydimethylsiloxane) except for proline and threonine exhibiting an inverted elution order [158].

In complexation enantio-GC, a rather consistent correlation between absolute configuration and elution order of 2-alkyl- and 2,3-dialkyl-substituted oxiranes on metal(II)-*bis*(perfluoroacyl-(1R)-camphorates) (Figure 4.6) was found to lead to the formulation of a quadrant rule [228,229]. However, also striking exceptions to the rule were noted for structurally related racemates. Thus, *trans*-(2S,3S)-dimethyloxirane (a "hard" donor selectand) and *trans*-(2R,3R)-dimethylthiirane (a "soft" donor selectand) were both eluted as the second peak on nickel(II) *bis*[3-heptafluorobutanoyl-(1R)-camphorate] despite their opposite configurations, whereas (R)-methyloxirane and (S)-methylthiirane showed the expected opposite elution order on the same selector [65]. (S)-2-*methyl*-tetrahydrofuran was eluted as the first peak, while (S)-2-*ethyl*-tetrahydrofuran was eluted as the second peak from nickel(II) *bis*[3-(heptafluorobutanoyl)-(1R)-camphorate] (Figure 4.6) [230]. These examples vividly showed that predictions of configurations from retention behavior for homologues or structurally related compounds may be ambiguous. Moreover, the temperature-dependent reversal of the elution order may obscure assignments of configurations by enantio-GC as the result of enthalpy–entropy compensation within an extended temperature range of operation (cf. Section 4.10). Thus, isopropyloxirane as compared to other alkyloxiranes exhibited a low isoenantioselective temperature T_{iso} on nickel(II) chelate selectors [5,72,81]. A peak reversal occurs above T_{iso}, and a nonconsistent elution order would result between isopropyloxirane and other homologues alkyloxiranes with identical absolute configuration at elevated temperatures in the enantio-GC experiment on simple thermodynamic grounds.

The dextrorotatory inhalation anesthetics isoflurane and desflurane (2-(difluoro-methoxy)-1,1,1,2-tetrafluoroethane), which differ only by chlorine versus fluorine at the stereogenic center at carbon, were eluted on the CSP Lipodex E before the levo-rotatory enantiomers (Figure 4.15) [9,231]. Thermodynamic considerations using the retention-increment R' approach supported the reasonable assumption that the enantiomers of isoflurane and desflurane with the same configuration possess the same sign of optical rotation and the same elution order on Lipodex E. The assignment of absolute configuration to dextrorotatory and levorotatory isoflurane and desflurane was carried out by vibrational circular dichroism (VCD) spectra. It was concluded that (+)-isoflurane had the (S)-configuration [232], whereas (+)-desflurane had the (R)-configuration [233]. This assignment was not in agreement with the expected elution order in enantio-GC, chemical shift differences in ^{1}H-NMR, and the substitution of chlorine by fluorine in the conversion of isoflurane to desflurane proceeding as an S_{N2}-inversion reaction. Later, an anomalous X-ray study [234] confirmed the revision of the desflurane configuration as (+)-(S) [235]. The case demonstrates that the interplay of different methods can clarify controversial configurational assignments, and that enantio-GC plays an important role in it [9].

4.8   THE METHOD OF ENANTIOMER LABELING IN ENANTIO-GC

The enantiomer of opposite configuration represents an ideal internal standard for the quantitation of the target enantiomer in a mixture [9]. This was first demonstrated in the analysis of chiral α-amino acids. Thus, the amount of an L-amino acid in a sample has been extrapolated from the change of the enantiomeric ratio after addition of a known amount of the oppositely configured D-amino acid (or the DL-racemate) employed as an internal label. This method was first conceived by Bonner [236,237] calling the internal standard an enantiomer marker. The approach, which was subsequently termed enantiomer labeling [238–240], is intriguing because enantiomers possess identical (nonchiroptical) properties in an achiral environment; therefore, the enantiomeric composition is not influenced by sample manipulation (isolation, derivatization, fractionation, storage) or by chromatographic manipulations (dilution, partitioning, splitting, injection, detection). Not even thermal or catalytic decomposition, losses, or incomplete isolation will obscure the analytical result. Thus, the added enantiomer serves as an overall internal standard through the whole analytical procedure. The method relies on the absence of self-recognition between enantiomers by molecular association (e.g., dimerization) in concentrated nonracemic mixtures (the *EE* effect) [241]. However, such nonlinear effects in enantioselective chromatography [242] constitute a rather rare phenomenon and can probably be totally ignored in diluted systems.

The amount of a particular enantiomer X_a present in the sample is calculated from the ratio of the peak areas of the (R)-enantiomer A_R and the enantiomeric (S)-label A_S multiplied by the amount of the (S)-label m_S added as the internal standard [238]:

$$X_a = m_S(A_R/A_S) \tag{4.1}$$

Substance-specific calibration factors f need not be considered by the enantiomer labeling method [238]. The method of enantiomer labeling can also be used for chiral compounds in samples and standards possessing only incomplete enantiomeric purities, including even racemic compositions. The method only requires the precise knowledge of the enantiomeric ratios of the sample and the standard that are readily accessible by the same enantio-GC method. The amount of the chiral component in a sample after addition of the chiral standard can be obtained by the general equation [238]

$$X_i = m_S[(A_R - A_S \cdot C_S)(1 + C_R)/(A_S - A_R \cdot C_R)(1 + C_S)] \tag{4.2}$$

where A_R = peak area of the (R)-enantiomer after addition of the standard
A_S = peak area of the (S)-enantiomer after addition of the standard
C_R = enantiomeric ratio $(S)/(R)$ of the sample
C_S = enantiomeric ratio $(R)/(S)$ of the standard
m_S = amount of enantiomeric standard (S) added
X_i = amount of the chiral component i (as sum of its enantiomers) present in the sample

Also, the amount of a racemate [(S) + (R)] present in a complex matrix can be quantitatively determined via the enantiomer-labeling method when a known amount of a single enantiomer [(S) or (R)] is added to the sample, and the change of the peak areas is determined by enantio-GC.

The quantitative determination of the inhalation anesthetic isoflurane (Figure 4.15, top) in blood samples during and after surgery has been performed by enantio-headspace GC-MS, employing enantiopure (+)-(S)-isoflurane as an internal standard [9,243]. The method of enantiomer labeling requires two measurements on the enantio-GC setup. The enantiomer-labeling method was also compared with the common internal standard method [9,243]. As internal standard, the racemic anesthetic enflurane (2-chloro-1-(difluoromethoxy)-1,1,2-trifluoroethane) (Figure 4.15, top) has been selected. Isomeric enflurane possesses the same molecular weight as isoflurane. The method of internal standardization requires only a single measurement on the enantio-GC setup in contrast to the enantiomer labeling method.

4.9 THERMODYNAMIC PARAMETERS OF CHIRALITY RECOGNITION IN ENANTIO-GC: ISOENANTIOSELECTIVE TEMPERATURE T_{iso}

The separation of enantiomers on a CSP by enantio-GC is governed by thermodynamics, i.e., by the different stabilities of the transient diastereomeric associates AD and AL formed between the enantiopure selector A present in the stationary phase and the selectand enantiomers D and L (D is arbitrarily eluted as the second peak in enantio-GC). The association equilibrium must be fast and reversible. The association constants K_{AD}^{assoc} and K_{AL}^{assoc} quantify the formation of energetically distinct diastereomeric associates. The ratio of the association constants is defined as the true enantioseparation factor α^{assoc}:

$$\alpha^{assoc} = K_{AD}^{assoc}/K_{AL}^{assoc} \tag{4.3}$$

Thus, α^{assoc} quantifies the true enantioselectivity imparted by the chiral selector A on the selectand enantiomers D and L [244]. Equation (4.3) is relevant for an undiluted selector A employed as a CSP [245]. However, in contemporary enantio-GC, chiral selectors (amino acid derivatives, metal complexes, or modified CDs) are mostly diluted in, or chemically bonded to, polysiloxanes used as achiral solvent matrix S. In the case of a diluted selector A present in S, the apparent enantioseparation factor α^{app} is defined as the ratio of the retention factors k_D and k_L of the selectand enantiomers D and L observed on the total CSP (A in S) [246]:

$$\alpha^{app} = k_D/k_L \tag{4.4}$$

Equation (4.4) is frequently used as a practical measure of the apparent enantioselectivity. Since the retention factors of the enantiomers D and L also include equivalent nonenantioselective contributions to retention arising from the presence of the achiral solvent S, α^{app} is always lower than α^{assoc}. The true enantioseparation factor α_{assoc} is obtained from the retention-increment R', whereby the achiral contribution to retention is separated from the chiral contribution to retention [247]. The calculation of R' requires (a) the determination of the retention factors $k°$ of the enantiomers D and L on a reference column containing only the pure solvent S and (b) the determination of the retention factors k of the enantiomers on the column containing the selector A diluted in, or chemically bonded to, the solvent S. On the achiral reference column, $k°$ is identical for the enantiomers. On the column containing the chiral selector A diluted in the solvent S, enantioselectivity is introduced into the separation process according to the different diastereomeric association equilibria: $A + D = AD$ (K_{AD}^{assoc}) and $A + L = AL$ (K_{AL}^{assoc}). Both columns must have the same dimensions, dead volumes, and coating parameters. The following equation applicable to each enantiomer has been derived (K^{assoc} is the thermodynamic association constant, and a_A is the activity of A in S) [5]:

$$R' = (k/k°) - 1 = K^{assoc} \cdot a_A \tag{4.5}$$

R' defines the ratio of the associated and free enantiomer in the stationary phase (A in S). Thus, $1/(1 + R')$ is the fraction of the enantiomer dissolved in the solvent S, and $R'/(1+R')$ is the fraction of the enantiomer associated with the selector A. For $R' = 1$, only half of the enantiomers present in the stationary phase are associated with the selector A. To become independent of all column parameters (except the temperature), relative retentions r_D, r_L, and $r°$ ($r = k/k_{ref} = t_R'/t_{R\,ref}'$), which are correlated to an inert reference standard, have been used in practice [5,66,247,248].

$$R' = (r/r°) - 1 = K^{assoc} \cdot a_A \tag{4.6}$$

Thus, for both selectand enantiomers D and L, the ratio $K_{AD}^{assoc}/K_{AL}^{assoc}$ is directly related to the ratio of the retention increments R'_D/R'_L, which is readily accessible from the relative retentions r_D, r_L, and $r°$, and the true enantioselectivity $-\Delta_{D,L}\Delta G^{true}$ in diluted selector systems is then obtained as follows:

$$-\Delta_{D,L}\Delta G^{true} = -\Delta_{D,L}\Delta H^{true} + T\Delta_{D,L}\Delta S^{true} = RT \ln \alpha^{assoc}$$
$$= RT \ln(K_{AD}^{assoc}/K_{AL}^{assoc}) = RT \ln(R'_D/R'_L) = RT \ln(r_D - r°/r_L - r°) \tag{4.7}$$

Equation (4.7) links the true enantioselectivity $-\Delta_{D,L}\Delta G^{true}$ (and its temperature dependence furnishing the enthalpic and entropic parameters $\Delta_{D,L}\Delta H^{true}$ and $\Delta_{D,L}\Delta S^{true}$) with the different retention increments R'_i of the enantiomers. From a thermodynamic point of view, the distinction between α^{app} and α^{assoc} is mandatory. A schematic representation of the inherent differences between α^{app} and α^{assoc} is shown in Figure 4.16 [249].

It follows from Equation (4.7) that the equation $-\Delta_{D,L}\Delta G \{=\} RT \ln \alpha^{app}$, frequently used in the literature to describe the true enantioselectivity, is not valid when diluted selectors A are employed in enantio-GC. Instead, Equation (4.8) is obtained by substituting k_D and k_L in Equation (4.4) by Equation (4.5):

$$\alpha^{app} = (K_{AD}^{assoc} a_A + 1)/(K_{AL}^{assoc} a_A + 1) = (R'_D + 1)/(R'_L + 1) \tag{4.8}$$

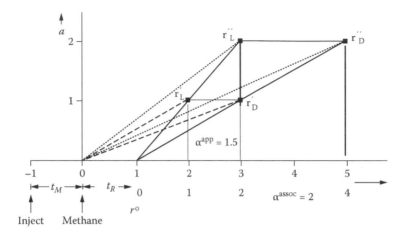

FIGURE 4.16 Schematic representation of the concept of the retention increment R' in enantio-GC separating achiral and chiral contributions to retention of racemic selectands on nonracemic chiral selectors diluted in a solvent.

α^{app} approaches α^{assoc} only when the association constants K^{assoc} or the activity a of A in S are high, resulting in $R'_i \gg 1$ [129] or when theoretically (but not practically) r° is zero (when nonenantioselective contributions to retention are negligible).

Moreover, α^{assoc}, and hence the true enantioselectivity $-\Delta_{D,L}\Delta G^{true} = RT \ln \alpha^{assoc}$, is independent of the concentration of A in S, whereas α^{app} is concentration dependent due a mixed retention mechanism [129,246]. However, a concentration dependence is not compatible with a thermodynamic quantity. This has been corroborated for the enantioseparation of methyl lactate and Lipodex E, whereby only the retention increment R' method afforded consistent thermodynamic results [128]. Due to their temperature dependence, values for α^{app} and α^{assoc} should only be quoted for isothermal chromatographic investigations and not for temperature-programmed measurements in enantio-GC.

The differentiation of apparent and true enantioselectivity in enantioseparations was at first advanced in enantioselective complexation GC utilizing metal chelates (Figure 4.6) [66] and was later extended to enantioselective inclusion GC utilizing modified CD selectors [129,250,251]. The inhalation anesthetics enflurane, isoflurane, and desflurane as well as a decomposition product of sevoflurane "compound B" [2-(fluoromethoxy)-3-methoxy-1,1,3,3-pentafluoropropane] show very large enantioseparation factors α on Lipodex E diluted in polysiloxanes (Figure 4.15, bottom) [249,250]. All thermodynamic data determined were independent of two different selector concentrations and the use of four different reference standards (n-alkanes, diethylether) [9,249,250]. The retention increment method has been used by Grinberg et al. in a thermodynamic study of N-trifluoro-O-alkyl nipecotic acid ester enantiomers on Chirasil-β-Dex [251]. The reliability of the retention increment R' approach was also confirmed for the separation of two enantiomeric pairs of the pheromone chalcogran (E- and Z-2-ethyl-1,6-dioxaspiro[4.4]nonane) between 80 and 120°C on a 25 m × 0.25 mm ID fused-silica column coated with 0.1 molal nickel(II) *bis* [3-(heptafluorobutanoyl)-(1R)-camphorate] [252].

As enantiomers cannot be distinguished on the reference column containing the achiral solvent matrix S, $r°_i$ is identical for the enantiomers D and L of a chiral compound. $r°_i$ need not be determined separately but can be extrapolated from two sets of data of the relative retention r_i of the D and L enantiomers at two (arbitrary) activities a_i of the chiral selector A in the solvent S of the columns (I) and (II) as a consequence of the following expression [252]:

$$r° = (r_L^{(I)} r_D^{(II)} - r_D^{(I)} r_L^{(II)}) / (r_L^{(I)} + r_D^{(II)}) - (r_D^{(I)} + r_L^{(II)}). \qquad (4.9)$$

This expression, which directly follows the theorem of intersecting lines of Thales (Figure 4.16), can be used to assess the nonenantioselective contributions to retention when $r°$ is not readily accessible due to the unavailability of the solvent S to prepare a reference column (e.g., for Chirasil-type CSPs). It suffices to collect retention data from two columns of different activities (or concentrations in dilute systems) of the selector A in S. For E-chalcogran and the aforementioned nickel(II) selector, the extrapolated and measured $r°$ value is almost identical (0.475 vs. 0.476 ± 0.001) [252]. Extrapolated values for $r°$ of compound B on Lipodex E were obtained using two columns coated with the CSP in polysiloxane PS-255 (5% and 10%, w/w) with four n-alkane reference standards (C5–C8) at 11 temperatures, and a satisfactory agreement between measured and extrapolated values of $r°$ was obtained [249]. Similarly, precise $r°$ data were extrapolated for enflurane, isoflurane, and desflurane (Figure 4.15, top) on two concentrations of Lipodex E in SE-54 using four reference standards [250]. These findings reinforced the validity of the retention increment R' approach, which relies on some experimental conditions, that is, use of traces of the selectand (10^{-8} g) to establish a true 1:1 association equilibrium and the presence of a dilute solution of the selector A in the solvent S (typically 0.05–0.1 molal).

It should be recognized that the enantioselectivity, as expressed by $-\Delta_{D,L}\Delta G^{true}$, is only determined by the *ratio* of the association constants of the selectand enantiomer D and L and the chiral selector A, $RT \ln(K_{AD}^{assoc}/K_{AL}^{assoc})$, regardless of whether the association interaction $-\Delta G_i = RT \ln K_i^{assoc}$ is weak, intermediate, or strong. Thus, high enantioselectivities are often observed already at low retention increments R', while low enantioselectivities may occur despite high retention increments R' at a strong overall molecular association. Consequently, the value of a chirality recognition factor $\chi = -\Delta_{D,L}\Delta G/-\Delta G_i$ is unpredictable, and it varies at random.

The measured true enantioselectivity $-\Delta_{D,L}\Delta G^{true}$ should be the same when an undiluted selector or a diluted selector [and calculated by Equation (4.7)] is used. For the selector mono-2-hydroxy-permethylated β-CD, differences of T_{iso} (Section 4.10) for 5-methyl-5-alkyl hydantoin derivatives were observed when the undiluted selector and the selector diluted in OV-1701 were compared [112]. Indeed, the retention increment R' method is only valid when dilute solutions of the selector A in the solvent S are employed [252]. In concentrated solutions, polysiloxane–selector interactions and a "salt-out effect" [5] cannot be neglected. Armstrong et al. proposed therefore a three-phase model for enantio-GC comprised of a permethylated CD/polysiloxane CSP [253].

4.10   ENTHALPY–ENTROPY COMPENSATION

The Gibbs–Helmholtz parameters $-\Delta_{D,L}\Delta H$ and $\Delta_{D,L}\Delta S$ of the thermodynamic enantioselectivity are readily accessible by linear van't Hoff plots when measurements are performed at different temperatures T according to

$$R \ln (R'_D/R'_L) = -\Delta_{D,L}\Delta G/T = -\Delta_{D,L}\Delta H/T + \Delta_{D,L}\Delta S. \qquad (4.10)$$

According to the Gibbs–Helmholtz equation, Equation (4.7), the true enantioselectivity $-\Delta_{D,L}\Delta G$ is governed by an enthalpy term $-\Delta_{D,L}\Delta H$ and an entropy term $\Delta_{D,L}\Delta S$ (the latter term is linked with the temperature T). For a 1:1 association equilibrium, both quantities obviously oppose each other in determining $-\Delta_{D,L}\Delta G$. Thus, enthalpy/entropy compensation arises by the fact that the more tightly bonded complex ($-\Delta H_D > -\Delta H_L$) is more ordered ($\Delta S_D < \Delta S_L$). Since the entropy term increases with temperature T, an isoenantioselective temperature will be reached at the compensation temperature $T_{iso} = \Delta_{D,L}\Delta H/\Delta_{D,L}\Delta S$ at which $\Delta_{D,L}\Delta G$ is zero (no enantioselectivity) since $K_D^{assoc} = K_L^{assoc}$. Peak coalescence (no enantioseparation) takes place at T_{iso} [254]. Above T_{iso}, enantioseparation commences again but with the inversion of the elution order. Below T_{iso}, enantiomeric separation is governed by the predominant enthalpic contribution to chirality recognition, whereas above T_{iso}, it is governed by the predominant entropic contribution to chirality recognition with the stronger bonded enantiomer D ($-\Delta H_D > -\Delta H_L$) eluted as the first peak. Whereas the sign of the enantioselectivity changes at T_{iso} when going from low to high temperatures, the association constants K_i^{assoc} between the selectand enantiomers D and L and selector A steadily decrease with increasing temperature T. The existence of an isoenantioselective temperature T_{iso} in enantio-GC has been observed independently in hydrogen-bonding equilibria [244,255] and in metal complexation equilibria [256,257], whereby the peak reversal for the enantiomers below and above T_{iso} was clearly established in the gas chromatograms. Temperature-dependent reversals of the elution order of the enantiomer of alanine-ECPA and valine-ECPA (ECPA = N-ethoxycarbonyl n-propylamide) by enantio-GC were observed on a 20 m × 0.25 mm (ID) fused-silica capillary coated with 0.25 μm on Chirasil-L-Val-C$_{11}$ between 100 and 150°C and 100 kPa (100–110°C) and 50 kPa (120–150°C) dihydrogen (Figure 4.17). Isoenantioselective temperatures T_{iso} for alanine-ECPA and valine-ECPA are 120 and 114°C, respectively [244]. Interestingly, the proline-ECPA devoid of an additional NH-bonding property exhibited a totally different van't Hoff plot [244]. This applied also for the proline trifluoroacetyl ethyl ester as compared to alanine-TFA-Et and valine-TFA-Et [244].

According to Figure 4.18 (top), the coordination interaction of isopropyloxirane on nickel(II) bis[3-(heptafluorobutanoyl)-8-methylene-(1R)-camphorate] steadily decreases between 55 and 110°C, as seen from the drop of the retention time from 50 to 2.5 min. While no enantioseparation is observed between 70 and 90°C, the (R)-enantiomer is eluted after the (S)-enantiomer at 55°C. Above the coalescence region at T_{iso}, enantioseparation resumes again at 110°C, whereby the stronger bonded (R)-enantiomer is now eluted before the (S)-enantiomer due to entropy control of chirality recognition. As expected from the thermodynamic origin of the

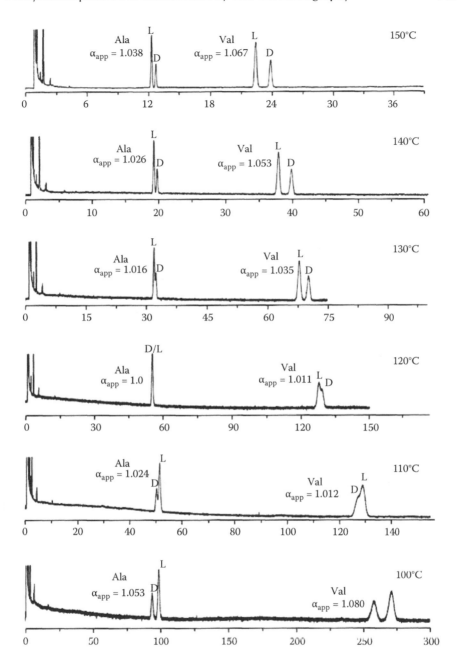

FIGURE 4.17 Temperature-dependent reversal of the elution order of the enantiomer of alanine-ECPA and valine-ECPA (ECPA = *N*-ethoxycarbonyl *n*-propylamide) by enantio-GC on Chirasil-L-Val-C$_{11}$ on a 20 m × 0.25 mm (ID) fused-silica capillary coated with 0.25-μm selector film thickness between 100 and 150°C and 50 kPa (120–150°C) and 100 kPa (100–110°C) dihydrogen. The α-amino acids were enriched with the L-enantiomer. Timescale: minutes. (From A. Levkin et al., *Anal. Chem. 79* (2007) 4401–4409 with permission. © American Chemical Society.)

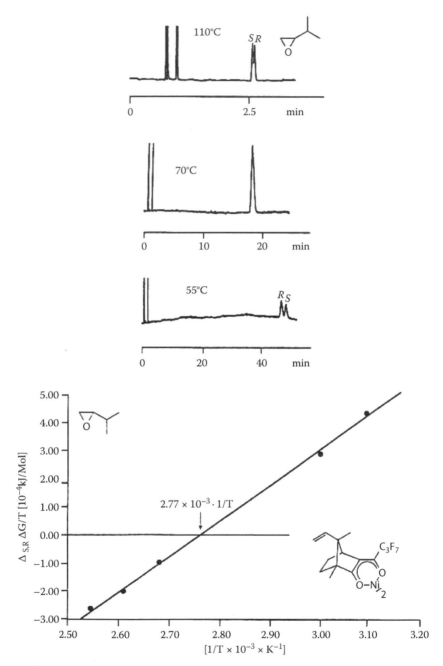

FIGURE 4.18 Top: Temperature-dependent reversal of the elution order of the enantiomers of isopropyloxirane by complexation enantio-GC on nickel(II) *bis*[3-(heptafluorobutanoyl)-8-methylene-(1*R*)-camphorate] on a 22.25 m × 0.25 mm (ID) glass capillary coated with 0.126 molal selector in polysiloxane OV-101 between 55 and 110°C and 1 bar dinitrogen. Bottom: Linear van't Hoff plot and determination of the isoenantioselective temperature T_{iso} of 89°C. (From V. Schurig and F. Betschinger, *Chem. Rev.* 92 (1992) 873–888 with permission.)

phenomenon, the van't Hoff plot in 5°C intervals between 55 and 110°C is strictly linear (Figure 4.18, bottom) furnishing $-\Delta_{D,L}\Delta H$ = 0.297 kcal/mol and $\Delta_{D,L}\Delta S$ = −0.82 cal/K·mol and T_{iso} = 362 K (~89°C) [72].

It was observed that minor structural modifications of the nickel(II) chelate in the camphor moiety and minor changes in the oxirane structure dramatically influenced T_{iso} [72]. For example, for the two enantiomeric pairs of *sec*-butyloxirane possessing two stereogenic centers, one enantiomeric pair showed a T_{iso} of 95°C, whereas the other enantiomeric pair exhibited a T_{iso} as high as 620°C [72], reinforcing the difficulty to rationalize chirality recognition mechanisms in complexation enantio-GC [5]. For the enantiomeric pairs of *E*- and *Z*-chalcogran (2-ethyl-1,6-dioxaspiro[4.4] nonane), only the *E*-enantiomeric pair showed an inversion of the elution order at T_{iso} = 82°C on the modified selector nickel(II) *bis*[3-(heptafluorobutanoyl)-(*1S*)-10-isobutylene-camphorate] (Figure 4.19) [5]. Whereas no peak reversal for the enantiomers of chalcogran was observed on the parent selector nickel(II) *bis*[3-(heptafluorobutanoyl)-(*1R*)-camphorate] (Figure 4.6) up to 120°C, the polymer attachment of the selector in Chirasil-nickel (Figure 4.4, middle, Me = Ni) reduced T_{iso} to 80°C [257].

Peak reversals have also been observed for isomenthone on octakis(2,6-di-O-*tert*-butyldimethylsilyl)-γ-cyclodextrin [258] and of the methyl esters of the herbicides mecoprop and dichlorprop on Lipodex E [259]. However, a report of a peak reversal for the enantiomers of methyl lactate on a modified CD selector (Lipodex E) [260] could not be ascertained [244], while temperature-induced inversions of the elution order have been observed for the enantiomers of *N*-trifluoroacetyl-α-amino acid ethyl esters on the selector Chirasil-Dex [244]. Thus, the derivatives of valine and leucine showed a reversal of the elution order on Chirasil-Dex below and above T_{iso} = 70°, and only a single peak was observed at the coalescence temperature. For the isoleucine derivative, the isoenantioselective temperature was as low as T_{iso} = 30° [244].

The enthalpy/entropy compensation effect described by Horváth et al. [261] has been investigated in enantio-GC employing modified CDs as selectors [145,262]. A related enthalpy/entropy compensation approach based on the plot of ln R' versus ΔH for the second eluted enantiomer has also been described by Grinberg et al. [251]. Enthalpy/entropy compensation must be considered for molecular modeling studies whereby the importance of entropy changes should be taken into account.

4.11 MINIATURIZATION IN ENANTIO-GC

Most enantioseparations by enantio-GC are governed by the enthalpy term of the Gibbs–Helmholtz equation, Equation (4.7). Consequently, enantioselectivity increases by reducing the elution temperature. As involatile racemates usually require a high elution temperature, it is advisable to use short columns (1–10 m × 0.25 mm ID [130,133,167,263] or just 1.5 m × 0.05 mm ID capillaries; Figure 4.11). The quest for miniaturization in enantio-GC has been addressed previously in detail [168]. The loss of efficiency arising from the smaller theoretical plate number N of a short column is often compensated by the gain of enantioselectivity due to the increased

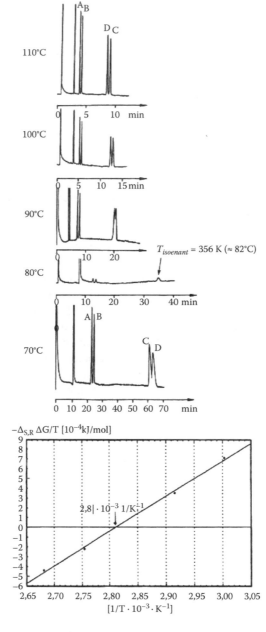

FIGURE 4.19 Top: Temperature-dependent reversal of the elution order of the enantiomers of E-chalcogran (C,D → D,C) by complexation enantio-GC on nickel(II) *bis*[3-(heptafluorobutanoyl)-(1S)-10-isobutylene-camphorate] on a 20.5 m × 0.25 mm (ID) glass capillary coated with 0.129 molal selector in polysiloxane OV-101 between 70 and 110°C and 1 bar dinitrogen [5]. Bottom: Linear van't Hoff plot and determination of the isoenantioselective temperature T_{iso} of 82°C. (From V. Schurig, *J. Chromatogr. A* 965 (2002) 315–356 with permission. © Elsevier.)

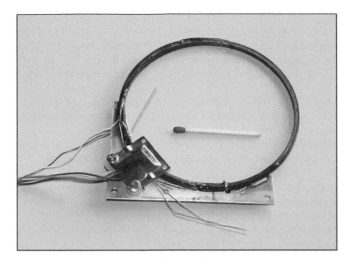

FIGURE 4.20 A prototype of a 10 m × 0.25mm (i.d.) fused silica capillary column coated with 0.25–0.125 μm CSP and connected to a micro-machined thermal conductivity detector as part of the COSAC enantio-GC experiment now present in outer space [156]. Courtesy, Prof. U. Meierhenrich.

enantioseparation factor α^{true} at the lower elution temperature. Three short 10 m × 0.25 mm ID fused-silica columns (Figure 4.20) coated with Chirasil-Val, Chirasil-Dex, or octakis(2,6-di-O-pentyl-3-O-trifluoroacetyl)-γ-cyclodextrin (DPTFA-γ-CD) are on board of the *Rosetta* spacecraft of European Space Agency (ESA) equipped with the Cometary Sampling and Composition (COSAC) system and heading toward the comet 67P/Churyumov-Gerasimenko and landing there in November 2014 in an attempt to search for extraterrestrial homochirality [156,264].

REFERENCES

1. V. Schurig, Emanuel Gil-Av and the separation of enantiomers on chiral stationary phases by chromatography, in: *Milestones in chromatography* (L. S. Ettre, Ed.), *LC·GC N. Am.* 25 (April 2007) 382–395.
2. L. He and T. E. Beesley, Applications of enantiomeric gas chromatography: a review, *J. Liq. Chromatogr. Relat. Technol.* 28 (2005) 1075–1114.
3. V. Schurig, Separation of enantiomers by gas chromatography on chiral stationary phases, in: *Chiral separation methods for pharmaceutical and biotechnological products (S. Ahuja, Ed.)*, Wiley, Hoboken, NJ, 2011, Chapter 9, pp. 251–297.
4. T. Beesley and R. E. Majors, The state of the art in chiral gas chromatography, *LC·GC N. Am.* 29 (August 1, 2011) 642–651 and *LC·GC Eur.* (May 1, 2012) 232–243.
5. V. Schurig, Practice and theory of enantioselective complexation gas chromatography, *J. Chromatogr. A* 965 (2002) 315–356.
6. V. Schurig, Use of derivatized cyclodextrins as chiral selectors for the separation of enantiomers by gas chromatography, *Ann. Pharmaceut. Française* 68 (2010) 82–98.

7. V. Schurig, Gas-chromatographic enantioseparation of derivatized α-amino acids on chiral stationary phases—past and present, *J. Chromatogr. B* 879 (2011) 3122–3140.
8. V. Schurig and D. Kreidler, Gas-chromatographic enantioseparation of unfunctionalized chiral hydrocarbons: an overview, in: *Chiral separations: Methods and protocols, 2nd edition* (G. K. E. Scriba, Ed.), Humana Press, Springer, New York, 2013, Chapter 3, pp. 45–67.
9. V. Schurig, Salient features of enantioselective gas chromatography: The enantiomeric differentiation of chiral inhalation anesthetics as a representative methodological case in point, *Top. Curr. Chem.* 340 (2013) 153–208.
10. W. Vetter and V. Schurig, Enantioselective determination of chiral organochlorine compounds in biota by gas chromatography on modified cyclodextrins, *J. Chromatogr. A* 774 (1997) 143–175.
11. W. Vetter and K. Bester, Gas chromatographic enantioseparation of chiral pollutants—techniques and results, in: *Chiral analysis* (K. W. Busch and M. A. Busch, Eds.), Elsevier, New York, 2006, Chapter 6, pp. 131–213.
12. H. Hühnerfuss and M. R. Shah, Enantioselective chromatography—a powerful tool for the determination of biotic and abiotic transformation processes of chiral environmental pollutants, *J. Chromatogr. A* 1216 (2009) 481–502.
13. V. Schurig, Preparative-scale separation of enantiomers on chiral stationary phases by gas chromatography, in: *Enantiomer separation: fundamentals and practical methods* (F. Toda, Ed.), Kluwer, Dordrecht, Netherlands, 2004 pp. 267–300.
14. E. Gil-Av and D. Nurok, Resolution of optical isomers by gas chromatography of diastereomers, in: *Advances in chromatography* (G. C. Giddings and R. A. Heller, Eds.), Decker, New York, 10, 1974, pp. 99–172.
15. V. Schurig, Terms for the quantitation of a mixture of stereoisomers, *Top. Curr. Chem.* 340 (2013) 21–40.
16. E. Gil-Av, Present status of enantiomeric analysis by gas chromatography, *J. Mol. Evol.* 6 (1975) 131–144.
17. C. H. Lochmüller, R. W. Souter, Chromatographic resolution of enantiomers, *J. Chromatogr.* 113 (1975) 283–302.
18. V. Schurig, Gas chromatographic separation of enantiomers on optically active metalcomplex-free stationary phases, *Angew. Chem. Int. Ed. Engl.* 23 (1984) 747–765.
19. W. A. König, *The practice of enantiomer separation by capillary gas chromatography*, Hüthig, Heidelberg, Germany, 1987.
20. W. A. König, Enantioselective gas chromatography, *Trends Anal. Chem.* 12 (1993) 130–137.
21. V. Schurig, Enantiomer separation by gas chromatography on chiral stationary phases, *J. Chromatogr. A* 666 (1994) 111–129.
22. Z. Juvancz, P. Petersson, Enantioselective gas chromatography, *J. Microcol. Sep.* 8 (1996) 99–114.
23. V. Schurig, Separation of enantiomers by gas chromatography, *J. Chromatogr. A* 906 (2001) 275–299.
24. V. Schurig, Chiral separations using gas chromatography, *Trends Anal. Chem.* 21 (2002) 647–661.
25. S. G. Allenmark, *Chromatographic enantioseparation: methods and applications*, 2nd rev. edition, Ellis Horwood, New York, 1991.
26. S. Allenmark and V. Schurig, Chromatography on chiral carriers, *J. Mater. Chem.* 7 (1997) 1955–1963.
27. P. Schreier, A. Bernreuther, and M. Huffer, *Analysis of chiral organic molecules*, de Gruyter, Berlin, 1995, pp. 132–233.
28. T. E. Beesley and R. P. W. Scott, *Chiral chromatography*, Wiley, New York, 1999.

29. E. Gil-Av, B. Feibush, and R. Charles-Sigler, Separation of enantiomers by gas liquid chromatography with an optically active stationary phase, *Tetrahedr. Lett.* 7 (1966) 1009–1015.
30. D. H. Desty, J. N. Haresnape, and B. H. F. Whyman, Construction of long lengths of coiled glass capillary, *Anal. Chem.* 32 (1960) 302–304.
31. E. Gil-Av, B. Feibush, and R. Charles-Sigler, Separation of enantiomer by gas-liquid chromatography with an optically active stationary phase, in: *Gas chromatography 1966* (A. B. Littlewood, Ed.), Institute of Petroleum, London, 1967, pp. 227–239 (discussion: pp. 254–257).
32. E. Gil-Av and B. Feibush, Resolution of enantiomers by gas liquid chromatography with optically active stationary phases. Separation on packed columns, *Tetrahedr. Lett.* 8 (1967) 3345–3347.
33. B. Feibush, Chiral separation of enantiomers via selector/selectand hydrogen bondings, *Chirality* 10 (1998) 382–395.
34. B. Feibush, Interaction between asymmetric solutes and solvents. N-Lauroyl-L-valyl-t-butylamide as stationary phase in gas liquid partition chromatography, *J. Chem. Soc. Chem. Commun.* (1971) 544–545.
35. C. H. Lochmüller and R. W. Souter, Direct gas chromatographic resolution of enantiomers on optically active mesophases. II. Effects of stationary phase structure on selectivity, *J. Chromatogr.* 88 (1974) 41–54.
36. H. Frank, G. J. Nicholson, and E. Bayer, Rapid gas chromatographic separation of amino acid enantiomers with a novel chiral stationary phase, *J. Chromatogr. Sci.* 15 (1977) 174–176.
37. H. Frank, Gas chromatography of enantiomers on chiral stationary phases, in: *Chirality and biological activity* (B. Holmstedt, H. Frank, and B. Testa, Eds.), Liss, New York, 1990, Chapter 3, pp. 33–54.
38. G. J. Nicholson, H. Frank, and E. Bayer, Glass capillary gas chromatography of amino acid enantiomers, *J. High Resolut. Chromatogr. Commun.* 2 (1979) 411–415.
39. G. Lai, G. Nicholson, and E. Bayer, Immobilization of Chirasil-Val on glass capillaries, *Chromatographia* 26 (1988) 229–233.
40. T. Saeed, P. Sandra, and M. Verzele, Synthesis and properties of a novel chiral stationary phase for the resolution of amino acid enantiomers, *J. Chromatogr.* 186 (1979) 611–618.
41. W. A. König and I. Benecke, Gas chromatographic separation of enantiomers of amines and amino alcohols on chiral stationary phases, *J. Chromatogr.* 209 (1981) 91–95.
42. B. Koppenhoefer, U. Mühleck, and K. Lohmiller, Backbone modification of Chirasil-Val I. Effect of loading on the separation of enantiomers by gas chromatography, *J. Chromatogr. A* 699 (1995) 215–221.
43. B. Koppenhoefer, U. Mühleck, M. Walser, and K. Lohmiller, Backbone modification of Chirasil-Val II. Introduction of a rigid cyclohexyl spacer, *J. Chromatogr. Sci.* 33 (1995) 217–222.
44. B. Koppenhoefer, U. Mühleck, and K. Lohmiller, Backbone modification of Chirasil-Val: effect of nonpolar side chains on enantiomer separation in gas chromatography, *Chromatographia* 40 (1995) 718–723.
45. P. A. Levkin, A. Levkina, and V. Schurig, Combining the enantioselectivities of L-valine diamide and permethylated β-cyclodextrin in one gas chromatographic chiral stationary phase, *Anal. Chem.* 78 (2006) 5143–5148.
46. H. Frank, I. Abe, and G. Fabian, A versatile approach to the reproducible synthesis of functionalized polysiloxane stationary phases, *J. High Resolut. Chromatogr.* 15 (1992) 444–448.
47. I. Abe, K.Terada, T. Nakahara, and H. Frank, New stereoselective GC phases: immobilized chiral polysiloxanes with (S)-(-)-t-leucine derivatives as selectors, *J. High Resolut. Chromatogr.* 21 (1998) 592–596.

48. N. Ôi, H. Kitahara, Y. Matsushita, and N. Kisu, Enantiomer separation by gas and high-performance liquid chromatography with tripeptide derivatives as chiral stationary phases, *J. Chromatogr. A* 722 (1996) 229–232.

49. M. Li, J. Huang, and T. Li, Enantiomeric separation of alcohols on a proline chiral stationary phase by gas chromatography, *J. Chromatogr. A* 1191 (2008) 199–204.

50. I. Abe and S. Ohtani, Novel chiral selectors anchored on polydimethylsiloxane as stationary phases for separation of derivatized amino acid enantiomers by capillary gas chromatography, *J. Sep. Sci.* 29 (2006) 319–324.

51. J. Pfeiffer and V. Schurig, Enantiomer separation of amino acid derivatives on a new polymeric chiral resorc[4]arene stationary phase by capillary gas chromatography, *J. Chromatogr. A* 840 (1999) 145–150.

52. A. Ruderisch, J. Pfeiffer, and V. Schurig, Synthesis of an enantiomerically pure resorcarene with pendant L-valine residues and its attachment to a polysiloxane (Chirasil-Calix), *Tetrahedron Asymmetry* 12 (2001) 2025–2030.

53. F. Narumi, N. Iki, T. Suzuki, T. Onodera, and S. Miyano, Syntheses of chirally modified thiacalix[4]arenas with enantiomeric amines and their application to chiral stationary phases for gas chromatography, *Enantiomer* 5 (2000) 83–93.

54. W. A. Bonner, M. A. Van Dort, and J. Flores, Quantitative gas chromatographic analysis of leucine enantiomers. Comparative study, *Anal. Chem.* 46 (1974) 2104–2107.

55. S. Abdalla, E. Bayer, and H. Frank, Derivatives for separation of amino acid enantiomers, *Chromatographia* 23 (1987) 83–85.

56. P. Hušek, Rapid derivatization and gas chromatographic determination of amino acids, *J. Chromatogr.* 552 (1991) 289–299.

57. H. Zahradnickova, P. Hušek, and P. Simek, GC separation of amino acid enantiomers via derivatization with heptafluorobutyl chloroformate and Chirasil-L-Val column, *J. Sep. Sci.* 32 (2009) 3919–3924.

58. H. Kaspar, K. Dettmer, W. Gronwald, and P. J. Oefner, Automated GC–MS analysis of free amino acids in biological fluids, *J. Chromatogr. B* 870 (2008) 222–232.

59. M. C. Pietrogrande and G. Basaglia, Enantiomeric resolution of biomarkers in space analysis: chemical derivatization and signal processing for gas chromatography-mass spectrometry analysis of chiral amino acids, *J. Chromatogr. A* 1217 (2010) 1126–1133.

60. M. Junge, H. Huegel, and P. J. Marriott, Enantiomeric analysis of amino acids by using comprehensive two-dimensional gas chromatography, *Chirality* 19 (2007) 228–234.

61. V. Schurig, Separation of deuteriated ethylenes $C_2H_{4-n}D_n$ by complexation chromatography on a rhodium(I) complex, *Angew. Chem. Int. Ed.* 15 (1976) 304.

62. V. Schurig, Resolution of a chiral olefin by complexation chromatography on an optically active rhodium(I) complex, *Angew. Chem. Int. Ed.* 16 (1977) 110.

63. V. Schurig and E. Gil-Av, Chromatographic resolution of chiral olefins. Specific rotation of 3-methylcyclopentene and related compounds, *Isr. J. Chem.* 15 (1976/77) 96–98.

64. B. T. Golding, P. J. Sellars, and A. K. Wong, Resolution of racemic epoxides on G.L.C. columns containing optically active lanthanoid complexes, *J. Chem. Soc. Chem. Commun.* (1977) 570–571.

65. V. Schurig and W. Bürkle, Extending the scope of enantiomer resolution by complexation gas chromatography, *J. Am. Chem. Soc.* 104 (1982) 7573–7580.

66. V. Schurig and R. Weber, Manganese(II)-bis(3-heptafluorobutyryl-1R-camphorate): a versatile agent for the resolution of racemic cyclic ethers by complexation gas chromatography, *J. Chromatogr.* 217 (1981) 51–70.

67. V. Schurig, Resolution of enantiomers and isotopic compositions by selective complexation gas chromatography on metal complexes, *Chromatographia* 13 (1980) 263–270.

68. V. Schurig, Enantiomer analysis by complexation gas chromatography. Scope, merits and limitations, *J. Chromatogr.* 441 (1988) 135–153.

69. R. Weber and V. Schurig, Complexation gas chromatography—a valuable tool for the stereochemical analysis of pheromones, *Naturwissenschaften* 71 (1984) 408–413.

70. V. Schurig, Enantiomer separation by complexation gas chromatography—applications in chiral analysis of pheromones and flavours, in: *Bioflavour '87* (P. Schreier, Ed.), de Gruyter, Berlin, 1988, pp. 35–54.

71. V. Schurig and R. Weber, Use of glass and fused-silica open tubular columns for the separation of structural, configurational and optical isomers by selective complexation gas chromatography, *J. Chromatogr.* 289 (1984) 321–332.

72. V. Schurig and F. Betschinger, Metal-mediated enantioselective access to unfunctionalized aliphatic oxiranes: prochiral and chiral recognition, *Chem. Rev.* 92 (1992) 873–888.

73. D. Wistuba, H.-P. Nowotny, O. Träger, and V. Schurig, Cytochrome P-450-catalyzed asymmetric epoxidation of simple prochiral and chiral aliphatic alkenes: species dependence and effect of enzyme induction on enantioselective oxirane formation, *Chirality* 1 (1989) 127–136.

74. D. Wistuba, O. Träger (in part), and V. Schurig, Enantio- and regioselectivity in the epoxide-hydrolase-catalyzed ring opening of simple aliphatic oxiranes (part I and part II), *Chirality* 4 (1992) 178–192.

75. V. Schurig, W. Bürkle, K. Hintzer, and R. Weber, Evaluation of nickel(II) bis[α-(heptafluorobutanoyl)-terpeneketonates] as chiral stationary phases for the enantiomer separation of alkyl-substituted cyclic ethers by complexation gas chromatography, *J. Chromatogr.* 475 (1989) 23–44.

76. V. Schurig, D. Schmalzing, and M. Schleimer, Enantiomer separation on immobilized Chirasil-Metal and Chirasil-Dex by gas chromatography and supercritical fluid chromatography, *Angew. Chem. Int. Ed. Engl.* 30 (1991) 987–989.

77. M. J. Spallek, G. Storch, and O. Trapp, Straightforward synthesis of poly(dimethylsiloxane) phases with immobilized 3-(perfluoroalkanoyl)-(1R)-camphorate metal complexes and their application in enantioselective complexation gas chromatography, *Eur. J. Org. Chem.* 21 (2012) 3929–3945.

78. S. Stockinger, M. J. Spallek, and O. Trapp, Investigation of novel immobilized 3-(perfluoroalkanoyl)-(1R)-camphorate nickel complexes in enantioselective complexation gas chromatography, *J. Chromatogr. A* 1269 (2012) 346–351.

79. V. Schurig, Peak coalescence phenomena in enantioselective chromatography, *Chirality* 10 (1998) 140–146.

80. W. Bürkle, H. Karfunkel, and V. Schurig, Dynamic phenomena during enantiomer resolution by complexation gas chromatography. A kinetic study of enantiomerization, *J. Chromatogr.* 288 (1984) 1–14.

81. V. Schurig, Contributions to the theory and practice of the chromatographic separation of enantiomers, *Chirality* 17 (2005) S205–S226.

82. V. Schurig, M. Jung, and M. Schleimer, F. G. Klärner, Investigation of the enantiomerization barrier of homofuran by computer-simulation of interconversion profiles obtained by complexation gas chromatography, *Chem. Ber.-Recueil* 125 (1992) 1301–1303.

83. V. Schurig, F. Keller, S. Reich, and M. Fluck, Dynamic phenomena involving chiral dimethyl-2,3-pentadienedioate in enantioselective gas chromatography and NMR spectroscopy, *Tetrahedron Asymmetry* 8 (1997) 3475–3480.

84. O. Trapp, G. Schoetz, and V. Schurig, Determination of enantiomerization barriers by dynamic and stopped-flow chromatographic methods, *Chirality* 13 (2001) 403–414.

85. J. Krupčik, P. Oswald, P. Májek, P. Sandra, and D. W. Armstrong, Determination of the interconversion energy barrier of enantiomers by separation methods, *J. Chromatogr. A* 1000 (2003) 779–800.

86. O. Trapp, Interconversion of stereochemically labile enantiomers (enantiomerization), *Top. Curr. Chem.* 341 (2013) 231–270.
87. V. Schurig, Supramolecular chromatography, in: *Applications of supramolecular chemistry* (H.-J. Schneider, Ed.), CRC Press, Boca Raton, FL, 2012, Chapter 6, pp. 129–157.
88. S.-M. Xie, X.-H. Zhang, Z.-J. Zhang, M. Zhang, J. Jia, and L.-M. Yuan, A 3-D open-framework material with intrinsic chiral topology used as stationary phase in gas chromatography, *Anal. Bioanal. Chem.* 405 (2013) 3407–3412.
89. J. Snopek, E. Smolková-Keulemansová, T. Cserháti, K. H. Gahm, and A. Stalcup, Cyclodextrins in analytical separation methods, in: *Comprehensive supramolecular chemistry* (J. Szejtli and T. Osa, Eds.), Vol. 3: *Cyclodextrins*, Pergamon, 1996, Chapter 18, pp. 516–571.
90. V. Schurig and H.-P. Nowotny, Gas chromatographic separation of enantiomers on cyclodextrin derivatives, *Angew. Chem. Int. Ed. Engl.* 29 (1990) 939–957.
91. W. A. König, G*as chromatographic enantiomer separation with modified cyclodextrins*, Hüthig, Heidelberg, Germany, 1992.
92. W. Li and T. M. Rossi, Derivatized cyclodextrins as chiral gas chromatographic stationary phases and their potential applications in the pharmaceutical industry, in: *The impact of stereochemistry on drug development and use* (H. Y. Aboul-Enein, and I. W. Wainer, Eds.), Vol. 142 in *Chemical Analysis* (J. D. Winefordner, Ed.), Wiley, New York, 1997, Chapter 15, pp. 415–436.
93. Z. Juvancz and J. Szejtli, The role of cyclodextrins in chiral selective chromatography, *Trends Anal. Chem.* 21 (2002) 379–388.
94. L. Li, M. Zi, C. X. Ren, and L. M. Yuan, The development of chiral stationary phases in gas chromatography, *Progr. Chem. (China)* 19 (2007) 393–403.
95. T. Kościelski, D. Sybilska, and J. Jurczak, Separation of α- and β-pinene into enantiomers in gas-liquid chromatography systems via α-cyclodextrin inclusion complexes, *J. Chromatogr.* 280 (1983) 131–134.
96. T. Kościelski, D. Sybilska, and J. Jurczak. New chromatographic method for the determination of the enantiomeric purity of terpenoic hydrocarbons, *J. Chromatogr.* 364 (1986) 299–303.
97. R. Ochocka, D. Sybilska, M. Aszemborska, J. Kowalczyk, and J. Goronowicz. Approach to direct chiral recognition of some terpenic hydrocarbon constituents of essential oils by gas chromatography systems via α-cyclodextrin complexation, *J. Chromatogr.* 543 (1991) 171–177.
98. M. Lindström, T. Norin, and J. Roeraade, Gas chromatographic separation of monoterpene hydrocarbon enantiomers on α-cyclodextrin, *J. Chromatogr.* 513 (1990) 315–320.
99. P. Sandra, Editorial: the enantioselectivity of derivatized cyclodextrins—a plea for more systematic studies, *J. High Resolut. Chromatogr.* 13 (1990) 665.
100. V. Schurig and H.-P. Nowotny, Separation of enantiomers on diluted permethylated β-cyclodextrin by high-resolution gas chromatography, *J. Chromatogr.* 441 (1988) 155–163.
101. H.-P. Nowotny, D. Schmalzing, D. Wistuba, and V. Schurig, Extending the scope of enantiomer separation on diluted methylated β-cyclodextrin derivatives by high-resolution gas chromatography, *J. High Resolut. Chromatogr.* 12 (1989) 383–393.
102. W. A. König, S. Lutz, P. Mischnick-Lübbecke, B. Brassat, and G. Wenz, Cyclodextrins as chiral stationary phases in capillary gas chromatography. I. Pentylated α-cyclodextrin, *J. Chromatogr.* 447 (1988) 193–197.
103. D. W. Armstrong, W. Li, C.-D. Chang, and J. Pitha, Polar-liquid, derivatized cyclodextrin stationary phases for the capillary gas chromatography separation of enantiomers, *Anal. Chem.* 62 (1990) 914–923.

104. W.-Y. Li, H. L. Jin, and D. W. Armstrong, 2,6-Di-*O*-pentyl-3-*O*-trifluoroacetyl cyclodextrin liquid stationary phases for capillary gas chromatographic separation of enantiomers, *J. Chromatogr.* 509 (1990) 303–324.

105. Z. Juvancz, G. Alexander, and J. Szejtli, Permethylated β-cyclodextrin as stationary phase in capillary gas chromatography, *J. High Resolut. Chromatogr.* 10 (1987) 105–107.

106. G. Alexander, Z. Juvancz, and J. Szejtli, Cyclodextrins and their derivatives as stationary phases in GC capillary columns, *J. High Resolut. Chromatogr.* 11 (1988) 110–113.

107. A. Venema and P. J. A. Tolsma, Enantiomer separation with capillary gas chromatography columns coated with cyclodextrins. Part I: Separation of enantiomeric 2-substituted propionic acid esters and some lower alcohols with permethylated β-cyclodextrin, *J. High Resolut. Chromatogr.* 12 (1989) 32–34.

108. S. Mayer, D. Schmalzing, M. Jung, and M. Schleimer, A chiral test mixture for permethylated β-cyclodextrin-polysiloxane gas-liquid chromatography phases: the Schurig test mixture, *LC·GC Int.* 5 (April 1992) 58–59.

109. W. Keim, A. Köhnes, W. Meltzow, and H. Römer, Enantiomer separation by gas chromatography on cyclodextrin chiral stationary phases, *J. High Resolut. Chromatogr.* 14 (1991) 507–529.

110. C. Bicchi, G. Artuffo, A. D'Amato, G. M. Nano, A. Galli, and M. Galli, Permethylated cyclodextrins in the GC separation of racemic mixtures of volatiles: Part 1, *J. High Resolut. Chromatogr.* 14 (1991) 301–305.

111. A. Jaus and M. Oehme, Consequences of variable purity of heptakis(2,3,6-tri-*O*-methyl)-β-cyclodextrin determined by liquid chromatography-mass spectrometry on the enantioselective separation of polychlorinated compounds, *J. Chromatogr. A* 905 (2001) 59–67.

112. H. Cousin, V. Peulon-Agasse, J.-C. Combret, and P. Cardinael, Mono-2,3 or 6-hydroxy methylated β-cyclodextrin (eicosa-*O*-methyl-β-cyclodextrin) isomers as chiral stationary phases for capillary GC, *Chromatographia* 69 (2009) 911–922.

113. W. A. König, Forum: collection of enantiomer separation factors obtained by capillary gas chromatography on chiral stationary phases, *J. High Resolut. Chromatogr.* 16 (1993) 312–323.

114. W. A. König, Forum: collection of enantiomer separation factors obtained by capillary gas chromatography on chiral stationary phases, *J. High Resolut. Chromatogr.* 16 (1993) 338–352.

115. W. A. König, Forum: collection of enantiomer separation factors obtained by capillary gas chromatography on chiral stationary phases, *J. High Resolut. Chromatogr.* 16 (1993) 569–586.

116. T. Beck, J. M. Liepe, J. Nandzik, and S. Rohn, A. Mosandl, Comparison of different di-tert-butyldimethyl-silylated cyclodextrins as chiral stationary phases in capillary gas chromatography, *J. High Resolut. Chromatogr.* 23 (2000) 569–575.

117. A. Shitangkoon and G. Vigh, Systematic modification of the separation selectivity of cyclodextrin-based gas chromatographic stationary phases by varying the size of the 6-O-substituents, *J. Chromatogr. A* 738 (1996) 31–42.

118. C. Bicchi, G. Cravotto, A. d'Amato, P. Rubiolo, A. Galli, and M. Galli, Cyclodextrin derivatives in gas chromatographic separation of racemates with different volatility. Part XV: 6-*O*-t-butyldimethylsilyl- versus 6-*O*-t-hexyldimethylsilyl β and γ derivatives, *J. Microcol. Sep.* 11 (1999) 487–500.

119. W. Blum and R. Aichholz, Gas chromatographic enantiomer separation on *tert*-butyldimethylsilylated β-cyclodextrin diluted in PS-086. A simple method to prepare enantioselective glass capillary columns, *J. High Resolut. Chromatogr.* 13 (1990) 515–518.

120. A. Dietrich, B. Maas, W. Messer, G. Bruche, V. Karl, A. Kaunzinger, and A. Mosandl, Stereoisomeric flavor compounds, part LVIII: the use of heptakis(2,3-di-*O*-methyl-6-*O*-tert-butyldimethylsilyl)-β-cyclodextrin as a chiral stationary phase in flavor analysis, *J. High Resolut. Chromatogr.* 15 (1992) 590–593.

121. C. Bicchi, D. D'Amato, V. Manzin, A. Galli, and M. Galli, (1996) Cyclodextrin derivatives in the gas chromatographic separation of racemic mixtures of volatile compounds. X. 2,3-Di-*O*-ethyl-6-*O*-tert-butyldimethylsilyl-β- and -γ-cyclodextrins, *J. Chromatogr. A* 742 (1996) 161–173.

122. A. Dietrich, B. Maas, V. Karl, P. Kreis, D. Lehmann, B. Weber, and A. Mosandl, Stereoisomeric flavor compounds part. LV: Stereodifferentiation of some chiral volatiles on heptakis(2,3-di-*O*-acetyl-6-*O*-tert-butyldimethylsilyl)-β-cyclodextrin, *J. High Resolut. Chromatogr.* 15 (1992) 176–179.

123. B. Maas, A. Dietrich, and A. Mosandl, Forum: collection of enantiomer separation factors obtained by capillary gas chromatography on chiral stationary phases, *J. High Resolut. Chromatogr.* 17 (1994) 109–115.

124. B. Maas, A. Dietrich, and A. Mosandl, Forum: collection of enantiomer separation factors obtained by capillary gas chromatography on chiral stationary phases, *J. High Resolut. Chromatogr.* 17 (1994) 169–173.

125. C. Bicchi, C. Cagliero, E. Liberto, B. Sgorbini, K. Martina, G. Cravotto, and P. Rubiolo, New asymmetrical per-substituted cyclodextrins (2-*O*-methyl-3-*O*-ethyl- and 2-*O*-ethyl-3-*O*-methyl-6-*O*-*t*-butyldimethylsilyl-β-derivatives) as chiral selectors for enantioselective gas chromatography in the flavour and fragrance field, *J. Chromatogr. A* 1217 (2010) 1106–1113.

126. C. Bicchi, C. Brunelli, G. Cravotto, P. Rubiolo, and M. Galli, Cyclodextrin derivatives in GC separation of racemates of different volatility. Part XVIII: 2-methyl-3-acetyl- and 2-acetyl-3-methyl-6-*O*-*t*-hexyldimethylsilyl-γ-cyclodextrin derivatives, *J. Sep. Sci.* 25 (2002) 125–134.

127. M. Junge and W. A. König, Selectivity tuning of cyclodextrin derivatives by specific substitution, *J. Sep. Sci.* 26 (2003) 1607–1614.

128. I. Špánic, J. Krupčik, and V. Schurig, Comparison of two methods for the gas chromatographic determination of thermodynamic parameters of enantioselectivity, *J. Chromatogr. A* 843 (1999) 123–128.

129. M. Jung, D. Schmalzing, and V. Schurig, Theoretical approach to the gas chromatographic separation of enantiomers on dissolved cyclodextrin derivatives, *J. Chromatogr.* 552 (1991) 43–57.

130. I. Hardt and W. A. König, Diluted versus undiluted cyclodextrin derivates in capillary gas chromatography and the effect of linear carrier gas velocity, column temperature, and length on enantiomer separation, *J. Microcol. Sep.* 5 (1993) 35–40.

131. V. Schurig, D. Schmalzing, U. Mühleck, M. Jung, M. Schleimer, P. Mussche, C. Duvekot, and J. C. Buyten, Gas chromatographic enantiomer separation on polysiloxane-anchored permethyl-β-cyclodextrin (Chirasil-Dex), *J. High Resolut. Chromatogr.* 13 (1990) 713–717.

132. P. Fischer, R. Aichholz, U. Bölz, M. Juza, and S. Krimmer, Permethyl-β-cyclodextrin, chemically bonded to polysiloxane: a chiral stationary phase with wider application range for enantiomer separation by capillary gas chromatography, *Angew. Chem. Int. Ed. Engl.* 29 (1990) 427–429.

133. V. Schurig, M. Jung, S. Mayer, M. Fluck, S. Negura, and H. Jakubetz, Unified enantioselective capillary chromatography on a Chirasil-DEX stationary phase. Advantages of column miniaturization, *J. Chromatogr. A* 694 (1995) 119–128.

134. H. Cousin, O. Trapp, V. Peulon-Agasse, X. Pannecoucke, L. Banspach, G. Trapp, Z. Jiang, J. C. Combret, and V. Schurig, Synthesis, NMR spectroscopic characterization and polysiloxane-based immobilization of the three regioisomeric monooctenylpermethyl-ß-cyclodextrins and their application in enantioselective GC, *Eur. J. Org. Chem.* (2003) 3273–3287.

135. H. Grosenick and V. Schurig, Enantioselective capillary gas chromatography and capillary supercritical fluid chromatography on an immobilized γ-cyclodextrin derivative, *J. Chromatogr.* 761 (1997) 181–193.

136. D. W. Armstrong, Y. Tang, T. Ward, and M. Nichols, Derivatized cyclodextrins immobilized on fused-silica capillaries for enantiomeric separations via capillary electrophoresis, gas chromatography, or supercritical fluid chromatography, *Anal. Chem.* 65 (1993) 1114–1117.

137. G. Yi, J. S. Bradshaw, B. E. Rossiter, A. Malik, W. Li, M. and L. Lee, New permethyl-substituted β-cyclodextrin polysiloxanes for use as chiral stationary phase in open tubular column chromatography, *J. Org. Chem.* 58 (1993) 4844–4850.

138. J. S. Bradshaw, Z. Chen, G. L. Yi, B. E. Rossiter, A. Malik, D. Pyo, H. Yun, D. R. Black, S. S. Zimmermann, M. L. Lee, W. D. Tong, and V. T. d'Souza, 6A,6B-β-Cyclodextrin-hexasiloxane copolymers: enantiomeric separations by a β-cyclodextrin-containing rotaxane copolymer, *Anal. Chem.* 67 (1995) 4437–4439.

139. M. Y. Nie, L. M. Zhou, Q. H. Wang, and D. Q. Zhu, Gas chromatographic enantiomer separation on single and mixed cyclodextrin derivative chiral stationary phases, *Chromatographia* 51 (2000) 736–740.

140. S. Tamogami, K.-I. Awano, M. Amaike, Y. Takagi, and T. Kitahara, Development of an efficient GLC system with a mixed chiral stationary phase and its application to the separation of optical isomers, *Flavour Fragr. J.* 16 (2001) 349–352.

141. M. Bayer and A. Mosandl, Improved gas chromatographic stereodifferentiation of chiral main constituents from different essential oils using a mixture of chiral stationary phases, *Flavour Fragr. J.* 19 (2004) 515–517.

142. S. Qi, P. Ai, C. Wang, L. Yuan, and G. Zhang, The characteristics of a mixed stationary phase containing permethylated-β-CD and perpentylated-β-CD in gas chromatography, *Sep. Purific. Technol.* 48 (2006) 310–313.

143. W. Vetter, K. Lehnert, and G. Hottinger, Enantioseparation of chiral organochlorines on permethylated β- and γ-cyclodextrin, as well as 1:1 mixtures of them, *J. Chromatogr. Sci.* 44 (2006) 596–601.

144. D. Kreidler, H. Czesla, and V. Schurig, A mixed stationary phase containing two versatile cyclodextrin-based selectors for the simultaneous gas chromatographic enantioseparation of racemic alkanes and racemic α-amino-acid derivatives, *J. Chromatogr. B* 875 (2008) 208–216.

145. A. Berthod, W. Li, and W. A. Armstrong, Multiple enantioselective retention mechanisms on derivatized cyclodextrin gas chromatographic chiral stationary phases, *Anal. Chem.* 64 (1992) 873–879.

146. G. Sicoli, Z. Jiang, L. Jicsinsky, and V. Schurig, Modified linear dextrins ("acyclodextrins") as new chiral selectors for the gas-chromatographic separation of enantiomers, *Angew. Chem. Int. Ed.* 44 (2005) 4092–4095.

147. G. Sicoli, F. Pertici, J. Jiang, L. Jicsinszky, and V. Schurig, Gas-chromatographic approach to probe the absence of molecular inclusion in enantioseparations by carbohydrates. Investigation of linear dextrins ("acyclodextrins") as novel chiral stationary phases, *Chirality* 19 (2007) 391–400.

148. D. W. Armstrong, L. He, and Y.-S. Liu, Examination of ionic liquids and their interaction with molecules, when used as stationary phases in gas chromatography, *Anal. Chem.* 71 (1999) 3873–3876.

149. J. Ding, T. Welton, and D. W. Armstrong, Chiral ionic liquids as stationary phases in gas chromatography, *Anal. Chem.* 76 (2004) 6819–6822.

150. T. Payagala and D.W. Armstrong, Chiral ionic liquids: a compendium of syntheses and applications (2005–2012), *Chirality* 24 (2012) 17–53.

151. A. Berthod, L. He, and D. W. Armstrong, Ionic liquids as stationary phase solvents for methylated cyclodextrins in gas chromatography, *Chromatographia* 53 (2001) 63–68.

152. K. Huang, X. Zhang, and D. W. Armstrong, Ionic cyclodextrins and ionic liquid matrices as chiral stationary phases for gas chromatography, *J. Chromatogr. A* 1217 (2010) 5261–5273.

153. M. Liang, M. Qi, C. Zhang, and R. Fu, Peralkylated-β-cyclodextrin used as gas chromatographic stationary phase prepared by sol-gel technology for capillary column, *J. Chromatogr. A* 1059 (2004) 111–119.
154. J. O. Grisales, P. J. Lebed, S. Keunchkarian, R. Francisco, F. R. González, and C. B. Castells, Permethylated β-cyclodextrin in liquid poly(oxyethylene) as a stationary phase for capillary gas chromatography, *J. Chromatogr. A* 1216 (2009) 6844–6851.
155. Y. Zhang, Z. S. Breitbach, C. L. Wang, and D. W. Armstrong, The use of cyclofructans as novel chiral selectors for gas chromatography, *Analyst* 135 (2009) 1076–1083.
156. F. Goesmann, H. Rosenbauer, R. Roll, C. Szopa, F. Raulin, R. Sternberg, G. Israel, U. Meierhenrich, W. Thiemann, and G. Munoz-Caro, COSAC, the cometary sampling and composition experiment on Philae, *Space Sci. Rev.* 128 (2007) 257–280.
157. U. J. Meierhenrich, *Amino acids and the asymmetry of life—with a foreword by Henri B. Kagan*, Springer, Heidelberg, Germany, 2008.
158. V. Schurig, M. Juza, M. Preschel, G. J. Nicholson, and E. Bayer, Gas-chromatographic enantiomer separation of proteinogenic amino acid derivatives: comparison of Chirasil-Val and Chirasil-γ-Dex used as chiral stationary phases, *Enantiomer* 4 (1999) 297–303.
159. W. H. Pirkle and C. J. Welch, Some thoughts on the coupling of dissimilar chiral columns or the mixing of chiral stationary phases for the separation of enantiomers, *J. Chromatogr. A* 731 (1996) 322–326.
160. P. A. Levkin and V. Schurig, Apparent and true enantioselectivity of single- and binary-selector chiral stationary phases in gas chromatography, *J. Chromatogr. A* 1184 (2008) 309–322.
161. A. Ruderisch, J. Pfeiffer, and V. Schurig, Mixed chiral stationary phase containing modified resorcinarene and β-cyclodextrin selectors bonded to a polysiloxane for enantioselective gas chromatography, *J. Chromatogr. A* 994 (2003) 127–135.
162. P. A. Levkin, A. Ruderisch, and V. Schurig, Combining the enantioselectivity of a cyclodextrin and a diamide selector in a mixed binary gas-chromatographic chiral stationary phase, *Chirality* 18 (2006) 49–63.
163. P. A. Levkin, A. Levkina, H. Czesla, S. Nazzi, and V. Schurig, Expanding the enantioselectivity of the gas-chromatographic chiral stationary phase Chirasil-Val-C-11 by doping it with octakis(3-*O*-butanoyl-2,6-di-*O*-n-pentyl)-γ-cyclodextrin, *J. Sep. Sci.* 30 (2007) 98–103.
164. G. Uccello-Barretta, S. Nazzi, S. Balzano, P. A. Levkin, V. Schurig, and P. Salvadori, Heptakis[2,3-di-*O*-methyl-6-*O*-(L-valine-tert-butylamide-N-alpha-ylcarbonylmethyl)]-ß-cyclodextrin: a new multifunctional cyclodextrin CSA for the NMR enantiodiscrimination of polar and apolar substrates, *Eur. Org. Chem.* (2007) 3219–3226.
165. O. Stephany, F. Dron, S. Tisse, A. Martinez, J.-M. Nuzillard, V. Peulon-Agasse, P. Cardinaël, and J.-P. Bouillon, (L)- or (D)-valine tert-butylamide grafted on permethylated β-cyclodextrin derivatives as new mixed binary chiral selectors. Versatile tools for capillary gas chromatographic enantioseparation, *J. Chromatogr. A* 1216 (2009) 4051–4062.
166. O. Stephany, S. Tisse, G. Coadou, J. P. Bouillon, V. Peulon-Agasse, and P. Cardinael, Influence of amino acid moiety accessibility on the chiral recognition of cyclodextrin-amino acid mixed selectors in enantioselective gas chromatography, *J. Chromatogr. A* 1270 (2012) 254–261.
167. M. Lindström, Improved enantiomer separation using very short capillary columns coated with permethylated β-cyclodextrin, *J. High Resolut. Chromatogr.* 14 (1991) 765–767.
168. V. Schurig and H. Czesla, Miniaturization of enantioselective gas chromatography, *Enantiomer* 6 (2001) 107–128.
169. H. Brückner and A. Schieber, Determination of free D-amino acids in mammalia by chiral gas chromatography-mass spectrometry, *J. High Resolut. Chromatogr.* 23 (2000) 576–582.

170. P. Chalier and J. Crouzet, Enantiodifferentiation of four γ-lactones produced by *Penicillium roqueforti, Chirality* 10 (1998) 786–790.
171. A. Glausch, J. Hahn, and V. Schurig, Enantioselective determination of chiral 2,2',3,3',4,6'-hexachloro-biphenyl (PCB 132) in human milk samples by multidimensional gas chromatography/electron capture detection and by mass spectrometry, *Chemosphere* 30 (1995) 2079–2085.
172. V. Schurig, M. Schleimer, M. Jung, S. Mayer, and A. Glausch, Enantiomer separation by GLC, SFC and CZE on high-resolution capillary columns coated with cyclodextrin derivatives, in: *Progress in flavour precursor studies* (P. Schreier, and P. Winterhalter, Eds.), Allured, Carol Stream, IL, 1993, pp. 63–75.
173. A. Berthod, X. Wang, K. H. Gahm, and D. W. Armstrong, Quantitative and stereoisomeric determination of light biomarkers in crude oil and coal samples, *Geochim. Cosmochim. Acta* 62 (1998) 1619–1630.
174. R. Schmidt, H. G. Wahl, H. Häberle, H.-J. Dieterich, and V. Schurig, Headspace gas chromatography-mass spectrometry analysis of isoflurane enantiomers in blood samples after anesthesia with the racemic mixture, *Chirality* 11 (1999) 206–211.
175. R. Liardon, S. Ledermann, and U. Ott, Determination of D-amino acids by deuterium labelling and selected ion monitoring, *J. Chromatogr.* 203 (1981) 385–395.
176. S. Weiner, Z. Kustanovich, and E. Gil-Av, Dead Sea scroll parchments: unfolding of the collagen molecules and racemization of aspartic acid, *Nature* 287 (1980) 820–823.
177. K. Nokihara and J. Gerhardt, Development of an improved automated gas-chromatographic chiral analysis system: application to non-natural amino acids and natural protein hydrolysates, *Chirality* 13 (2001) 431–434.
178. O. Tenberken, F. Worek, H. Thiermann, and G. Reiter, Development and validation of a sensitive gas-chromatography-ammonia chemical ionization mass spectrometry method for the determination of tabun enantiomers in hemolysed blood and plasma of different species. *J. Chromatogr. B* 878 (2010) 1290–1296.
179. H. P. Benschop, C. A. G. Konings, and L. P. A. De Jong, Gas chromatographic separation and identification of the four stereoisomers of 1,2,2-trimethylpropyl methylphosphonofluoridate (Soman). Stereospecificity of in vitro "detoxification" reactions, *J. Am. Chem. Soc.* 103 (1981) 4260–4262.
180. G. T. Eyres, S. Urban, P. D. Morrison, and P. J. Marriott, Application of microscale-preparative multidimensional gas chromatography with nuclear magnetic resonance spectroscopy for identification of pure methylnaphthalenes from crude oil, *J. Chromatogr. A* 1215 (2008) 168–176.
181. M. Kühnle, D. Kreidler, K. Holtin, H. Czesla, P. Schuler, V. Schurig, and K. Albert, Online coupling of enantioselective capillary gas chromatography with proton nuclear magnetic resonance spectroscopy, *Chirality* 22 (2010) 808–812.
182. M. Kühnle, D. Kreidler, H. Czesla, K. Holtin, P. Schuler, W. Schaal, V. Schurig, and K. Albert, On-line coupling of gas chromatography to nuclear magnetic resonance spectroscopy: method for the analysis of volatile stereoisomers, *Anal. Chem.* 80 (2008) 5481–5486.
183. G. Schomburg, H. Husmann, E. Hübinger, and W. A. König, Multidimensional capillary gas chromatography—enantiomeric separations of selected cuts using a chiral second column, *J. High Resolut. Chromatogr.* 7 (1984) 404–410.
184. L. Mondello, A. Verzera, P. Previti, F. Crispo, and G. Dugo, Multidimensional capillary GC-GC for the analysis of complex samples. 5. Enantiomeric distribution of monoterpene hydrocarbons and monoterpene alcohols, and linalyl acetate of bergamot (*Citrus bergamia* Risso et Poiteau) oils, *J. Agric. Food Chem.* 46 (1998) 4275–4282.
185. G. P. Blanch and J. Jauch, Enantiomeric composition of filbertone in hazelnuts in relation to extraction conditions. Multidimensional gas chromatography and gas chromatography/mass spectrometry in the single ion monitoring mode of a natural sample, *J. Agric. Food Chem.* 46 (1998) 4283–4286.

186. Y. Saritas, N. Bülow, C. Fricke, W. A. König, and H. Muhle, Sesquiterpene hydrocarbons in the liverwort *Dumortiera hirsuta, Phytochemistry* 48 (1998) 1019–1023.
187. C. Barba, G. Flores, and M. Herraiz, Stereodifferentiation of some chiral aroma compounds in wine using solid phase microextraction and multidimensional gas chromatography, *Food Chem.* 123 (2010) 846–851.
188. P. Rubiolo, B. Sgorbini, E. Liberto, C. Cordero, and C. Bicchi, Essential oils and volatiles: sample preparation and analysis, *Flavour Fragr. J.* 25 (2010) 282–290.
189. D. Sciarrone, L. Schipilliti, C. Ragonese, P. Q. Tranchida, P. Dugo, G. Dugo, and L. Mondello, Thorough evaluation of the validity of conventional enantio-gas chromatography in the analysis of volatile chiral compounds in mandarin essential oil: a comparative investigation with multidimensional gas chromatography, *J. Chromatogr. A* 1217 (2010) 1101–1105.
190. R. Shellie, L. Mondello, G. Dugo, and P. Marriott, Enantioselective gas chromatographic analysis of monoterpenes in essential oils of the family Myrtaceae, *Flavour Fragr. J.* 19 (2004) 582–585.
191. H.-E. Högberg, E. Hedenström, A.-B. Wassgren, M. Hjalmarsson, G. Bergström, J. Löfqvist, and T. Norin, Synthesis and gas chromatographic separation of the eight stereoisomers of diprionol and their acetates, components of the sex pheromone of pine sawflies, *Tetrahedron* 46 (1990) 3007–3018.
192. H. P. Benschop and L. P. A. De Jong, Toxicokinetics of soman: species variation and stereospecificity in elimination pathways, *Neurosci. Biobehav. Rev.* 15 (1991) 73–77.
193. A. Glausch, G. J. Nicholson, M. Fluck, and V. Schurig, Separation of the enantiomers of stable atropisomeric polychlorinated biphenyls (PCBs) by multidimensional gas chromatography on Chirasil-Dex, *J. High Resolut. Chromatogr.* 17 (1994) 347–349.
194. C. Bicchi, A. D'Amato, and P. Rubiolo, Review—cyclodextrin derivatives as chiral selectors for direct gas chromatographic separation of enantiomers in the essential oil, aroma and flavour field, *J. Chromatogr. A* 843 (1999) 99–121.
195. J. P. Marriott, R. Shellie, and C. Cornwell, Review—gas chromatographic technologies for the analysis of essential oils, *J. Chromatogr. A* 936 (2001) 1–22.
196. M. Kreck, A. Scharrer, S. Bilke, and A. Mosandl, Stir bar sorptive extraction(SBSE)-enantio-MDGC-MS—a rapid method for the enantioselective analysis of chiral flavour compounds in strawberries, *Eur. Food Res. Technol.* 213 (2001) 389–394.
197. M. Kreck, A. Scharrer, S. Bilke, and A. Mosandl, Enantioselective analysis of monoterpene compounds in essential oils by stir bar sorptive extraction (SBSE)-enantio-MDGC-MS, *Flavour Fragr. J.* 17 (2002) 32–40.
198. D. Sciarrone, C. Regonese, C. Carnovale, A. Piperno, P. Dugo, G. Dugo, and L. Mondello, Evaluation of tea tree oil quality and ascaridole: a deep study by means of chiral and multi heart-cuts multidimensional gas chromatography system coupled to mass spectrometry detection, *J. Chromatogr. A* 1217 (2010) 6422–6427.
199. P. J. Marriott, S. T. Chin, B. Maikhunthod, H. G. Schmarr, and S. Bieri, Multidimensional gas chromatography, *Trends Anal. Chem.* 34 (2012) 1–21.
200. P. Marriott and R. Shellie, Principles and applications of comprehensive two-dimensional gas chromatography, *Trends Anal. Chem.* 21 (2002) 573–583.
201. R. Shellie and P. J. Marriott, Comprehensive two-dimensional gas chromatography with fast enantioseparation, *Anal. Chem.* 74 (2002) 5426–5430.
202. M. Junge, H. Huegel, and P. J. Marriott, Enantiomeric analysis of amino acids by using comprehensive two-dimensional gas chromatography, *Chirality* 19 (2007) 228–234.
203. M. Junge, S. Bieri, H. Huegel, and P. J. Marriott, Fast comprehensive two-dimensional gas chromatography with cryogenic modulation, *Anal. Chem.* 79 (2007) 4448–4454.
204. P. Shellie, P. Marriott, and C. Cornwell, Application of comprehensive two-dimensional gas chromatography (GC×GC) to the enantioselective analysis of essential oils, *J. Sep. Sci.* 24 (2001) 823–830.

205. L. R. Bordajandi, P. Korytár, J. de Boer and M. J. Gonzáles, Enantiomeric separation of chiral polychlorinated biphenyls on β-cyclodextrin capillary columns by means of heart-cut multidimensional gas chromatography and comprehensive two-dimensional gas chromatography. Application to food samples, *J. Sep. Sci.* 28 (2005) 163–171.

206. V. Schurig and S. Reich, Determination of the rotational barriers of atropisomeric polychlorinated biphenyls (PCBs) by a novel stopped-flow multidimensional gas chromatographic technique, *Chirality* 10 (1998) 316–320.

207. S. Reich and V. Schurig, Stopped-flow multidimensional gas chromatography: a new method for the determination of enantiomerization barriers, *J. Microcol. Sep.* 11 (1999) 475–479.

208. S. Reich, O. Trapp, and V. Schurig, Enantioselective stopped-flow multidimensional gas chromatography—determination of the inversion barrier of 1-chloro-2,2-dimethylaziridine, *J. Chromatogr. A* 892 (2000) 487–498.

209. J. Krupčik, J. Mydlová, P. Májek, P. Šimon, and D. W. Armstrong, Methods for studying reaction kinetics in gas chromatography, exemplified by using the 1-chloro-2,2-dimethylaziridine interconversion reaction, *J. Chromatogr. A* 1186 (2008) 144–160.

210. F. de la Pena Moreno, G. P. Blanch, G. Flores, and M. L. R. del Castillo, Development of a method based on on-line reversed phase liquid chromatography and gas chromatography coupled by means of an adsorption-desorption interface for the analysis of selected chiral volatile compounds in methyl jasmonate treated strawberries, *J. Chromatogr. A* 1217 (2010) 1083–1088.

211. L. Mondello, G. Dugo, and K. D. Bartle, On-line microbore high performance liquid chromatography-capillary gas chromatography for food and water analyses, *J. Microcol. Sep.* 8 (1996) 275–310.

212. L. Mondello, G. Dugo, and K. D. Bartle, On-line HPLC-HRGC in the analytical chemistry of citrus essential oils, *Perfumer Flavorist* 21 (1996) 25–49.

213. M. L. R. del Castillo, E. G. Caballero, and M. Herraiz, Stereodifferentiation of chiral compounds using reversed-phase liquid chromatography coupled with capillary gas chromatography, *J. Chromatogr. Sci.* 41 (2003) 26–30.

214. C. Barba, R. M. Martínez, M. M. Calvo, G. Santa-María, and M. Herraiz, Chiral analysis by online coupling of reversed-phase liquid chromatography to gas chromatography and mass spectrometry, *Chirality* 24 (2012) 420–426.

215. A. Mosandl, Enantioselective capillary gas chromatography and stable isotope ratio mass spectrometry in the authenticity control of flavours and essential oils, *Food Rev. Int.* 11 (1995) 597–664.

216. U. Hener, W. Brand, A. Hilkert, D. Juchelka, and A. Mosandl, F. Podebrad, Simultaneous on-line analysis of $^{18}O/^{16}O$ and $^{13}C/^{12}C$ isotope ratios of organic compounds using GC-pyrolysis-IRMS, *Z. Lebensm. Unters. Forsch. A* 206 (1998) 230–232.

217. M. Greule, C. Hänsel, U. Bauermann, and A. Mosandl, Feed additives: authenticity assessment using multicomponent-/multielement-isotope ratio mass spectrometry, *Eur. Food Res. Technol.* 227 (2008) 767–776.

218. D. Juchelka, T. Beck, U. Hener, F. Dettmar, and A. Mosandl, Multidimensional gas chromatography coupled on-line with isotope ratio mass spectrometry (MDGC-IRMS): progress in the analytical authentication of genuine flavor components, *J. High Resolut. Chromatogr.* 21 (1998) 145–151.

219. S. Reichert, D. Fischer, S. Asche, and A. Mosandl, Stable isotope labelling in biosynthetic studies of dill ether, using enantioselective multidimensional gas chromatography, on-line coupled with isotope ratio mass spectrometry, *Flavour Fragr. J.* 15 (2000) 303–308.

220. B. Weckerle, R. Bastl-Borrmann, E. Richling, K. Hör, C. Ruff, and P. Schreier, Cactus pear (*Opuntia ficus* indica) flavour constituents—chiral evaluation (MDGC–MS) and isotope ratio (HRGC–IRMS) analysis, *Flavour Fragr. J.* 16 (2001) 360–363.

221. K. Schumacher, U. Hener, C. Patz, H. Dietrich, and A. Mosandl, Authenticity assessment of 2- and 3-methylbutanol using enantioselective and/or $^{13}C/^{12}C$ isotope ratio analysis, *Eur. Food Res. Technol.* 209 (1999) 12–15.

222. L. Schipilliti, P. Q. Tranchida, D. Sciarrone, M. Russo, P. Dugo, G. Dugo, and L. Mondello, Genuineness assessment of mandarin essential oils employing gas chromatography-combustion-isotope ratio MS (GC-C-IRMS), *J. Sep. Sci.* 33 (2010) 617–625.

223. L. Schipilliti, P. Dugo, I. Bonaccorsi, and L. Mondello, Headspace-solid phase microextraction coupled to gas chromatography-combustion-isotope ratio mass spectrometer and to enantioselective gas chromatography for strawberry flavoured food quality control, *J. Chromatogr. A* 1218 (2011) 7481–7486.

224. S. Pizzarello, Y. Huang, and M. Fuller, The carbon isotopic distribution of Murchison amino acids, *Geochim. Cosmochim. Acta* 23 (2004) 4963–4969.

225. P. Herwig, K. Zawatzky, M. Grieser, O. Heber, B. Jordon-Thaden, C. Krantz, O. Novotny, R. Repnow, V. Schurig, D. Schwalm, Z. Vager, A. Wolf, O. Trapp, and H. Kreckel, Imaging the absolute configuration of a chiral epoxide in the gas phase, *Science*, 342 (2013) 1084–1086.

226. H. D. Flack and G. Bernardinelli, The use of X-ray crystallography to determine absolute configuration, *Chirality* 20 (2008) 681–690.

227. P. L. Polavarapu, Molecular structure determination using chiroptical spectroscopy: where we may go wrong? *Chirality* 24 (2012) 909–920.

228. V. Schurig, B. Koppenhöfer, and W. Bürkle, Correlation of the absolute configuration of chiral epoxides by complexation chromatography; synthesis and enantiomeric purity of (+)- and (−)-1,2-epoxypropane, *Angew. Chem. Int. Ed. Engl.* 17 (1978) 937–939.

229. V. Schurig, Molecular recognition in complexation gas chromatography, in: *Chromatographic separations based on molecular recognition* (K. Jinno, Ed.), Wiley-VCH, Chapter 7, pp. 371–418.

230. K. Keinan, K. K. Seth, and R. Lamed, Organic synthesis with enzymes. 3. TBADH-catalyzed reduction of chloro ketones. Total synthesis of (+)-(S,S)-(cis-6-methyltetrahydropyran-2-yl)acetic acid: a civet constituent, *J. Am. Chem. Soc.* 108 (1986) 3474–3480.

231. V. Schurig, H. Grosenick, and M. Juza, Enantiomer separation of chiral inhalation anesthetics (enflurane, isoflurane and desflurane) by gas chromatography on a γ-cyclodextrin derivative, *Recl. Trav. Chim. Pays-Bas* 114 (1995) 211–219.

232. P. L. Polavarapu, A. L. Cholli, and G. Vernice, Absolute configuration of isoflurane, *J. Am. Chem. Soc.* 114 (1992) 10953–10955.

233. P. L. Polavarapu, A. L. Cholli, and G. Vernice, Determination of absolute configurations and predominant conformations of general inhalation anesthetics: desflurane, *J. Pharm. Sci.* 82 (1993) 791–793.

234. V. Schurig, M. Juza, B. S. Green, J. Horakh, and A. Simon, Absolute configuration of the inhalation anesthetics isoflurane and desflurane, *Angew. Chem. Int. Ed. Engl.* 35 (1996) 1680–1682.

235. P. L. Polavarapu, C. X. Zhao, A. L. Cholli, and G. G. Vernice, Vibrational circular dichroism, absolute configuration, and predominant conformations of volatile anesthetics: desflurane, *J. Phys. Chem.* 103 (1999) 6127–6132.

236. W. A. Bonner, Enantiomeric markers in the quantitative gas chromatographic analysis of optical isomers. Application to the estimation of amino acid degradation, *J. Chromatogr. Sci.* 11 (1973) 101–104.

237. N. E. Blair and W. A. Bonner, Quantitative determination of D ≠ L mixtures of optical enantiomers by gas chromatography, *J. Chromatogr.* 198 (1980) 185–187.

238. H. Frank, G. J. Nicholson, and E. Bayer, Enantiomer labelling, a method for the quantitative analysis of amino acids, *J. Chromatogr.* 167 (1978) 187–196.

239. H. Frank, A. Rettenmeier, H. Weicker, G. J. Nicholson, and E. Bayer, A new gas chromatographic method for determination of amino acid levels in human serum, *Clin. Chim. Acta* 105 (1980) 201–211.

240. E. Bayer, H. Frank, J. Gerhardt, and G. Nicholson, Capillary gas chromatographic analysis of amino acids by enantiomer labelling, *J. Assoc. Off. Anal. Chem.* 70 (1987) 234–240.

241. W.-L. Tsai, K. Hermann, E. Hug, B. Rohde, and A. S. Dreiding, Enantiomer-differentiation induced by an enantiomeric excess during chromatography with achiral phases, *Helv. Chim. Acta.* 68 (1985) 2238–2243.

242. O. Trapp and V. Schurig, Nonlinear effects in enantioselective chromatography: prediction of unusual elution profiles of enantiomers in non-racemic mixtures on an achiral stationary phase doped with small amounts of a chiral selector, *Tetrahedron Asymmetry* 21 (2010) 1334–1340.

243. M. Juza, H. Jakubetz, H. Hettesheimer, and V. Schurig, Quantitative determination of isoflurane enantiomers in blood samples during and after surgery via headspace gas chromatography-mass spectrometry, *J. Chromatogr. B* 735 (1999) 93–102.

244. A. Levkin, A. Levkina, H. Czesla, and V. Schurig, Temperature-induced inversion of the elution order of enantiomers in gas chromatography: *N*-ethoxycarbonyl propylamides and *N*-trifluoroacetyl ethyl esters of α-amino acids on Chirasil-Val-C11 and Chirasil-Dex stationary phases, *Anal. Chem.* 79 (2007) 4401–4409.

245. A. Shitangkoon, D. U. Staerk, and G. Vigh, Gas-chromatographic separation of the enantiomers of volatile fluoroether anesthetics using derivatized cyclodextrins stationary phases, Part 1, *J. Chromatogr. A* 657 (1993) 387–394.

246. P. A. Levkin and V. Schurig, Apparent and true enantioselectivity of single- and binary-selector chiral stationary phases in gas chromatography, *J. Chromatogr. A* 1184 (2008) 309–322.

247. V. Schurig, Review: elaborate treatment of retention in chemoselective chromatography—the retention increment approach and non-linear effects, *J. Chromatogr. A* 1216 (2009) 1723–1736.

248. V. Schurig, Relative stability constants of olefin-rhodium(II) vs. olefin-rhodium(I) coordination as determined by complexation gas chromatography, *Inorg. Chem.* 25 (1986) 945.

249. V. Schurig and R. Schmidt, Extraordinary chiral discrimination in inclusion gas chromatography. Thermodynamics of enantioselectivity between a racemic perfluorodiether and a modified γ-cyclodextrin, *J. Chromatogr. A* 1000 (2003) 311–324.

250. V. Schurig and M. Juza, Approach to the thermodynamics of enantiomer separation by gas chromatography-Enantioselectivity between the chiral inhalation anesthetics enflurane, isoflurane and desflurane and a diluted γ-cyclodextrin derivative, *J. Chromatogr. A* 757 (1997) 119–135.

251. N. T. McGachy, N. Grinberg, and N. Variankaval, Thermodynamic study of N-trifluoroacetyl-O-alkyl nipecotic acid ester enantiomers on diluted permethylated β-cyclodextrin stationary phase, *J. Chromatogr. A* 1064 (2005) 193–204.

252. V. Schurig and R. Link, Recent developments in enantiomer separation by complexation gas chromatography, in: *Chiral separations* (D. Stevenson and I. D. Wilson, Eds.), Plenum Press, New York, 1988, pp. 91–114.

253. V. Pino, A. W. Lantz, J. L. Anderson, A. Berthod, and D. W. Armstrong, Theory and use of the pseudophase model in gas-liquid chromatographic enantiomeric separations, *Anal. Chem.* 78 (2006) 113–119.

254. B. Koppenhoefer and E. Bayer, Chiral recognition in the resolution of enantiomers by GLC, *Chromatographia* 19 (1984) 123–130.

255. K. Watabe, R. Charles, and E. Gil-Av, Temperature dependent inversion of elution sequence in the resolution of α-amino acid enantiomers on chiral diamide selectors, *Angew. Chem. Int. Ed.* 28 (1989) 192–194.

256. V. Schurig, J. Ossig, and R. Link, Evidence for a temperature dependent reversal of the enantioselectivity in complexation gas chromatography on chiral phases, *Angew. Chem. Int. Ed.* 28 (1989) 194–196.

257. Z. Jiang and V. Schurig, Existence of a low isoenantioselective temperature in complexation gas chromatography. Profound change of enantioselectivity of a nickel(II) chiral selector either bonded to, or dissolved in, poly(dimethylsiloxane), *J. Chromatogr. A* 1186 (2008) 262–270.

258. B. Maas, A. Dietrich, T. Beck, S. Börner, and A. Mosandl, Di-tert-butyldimethylsilylated cyclodextrins as chiral stationary phases: thermodynamic investigations, *J. Microcol. Sep.* 7 (1995) 65–73.

259. W. A. König, D. Icheln, T. Runge, B. Pfaffenberger, P. Ludwig, and H. Hühnerfuss, Gas chromatographic enantiomer separation of agrochemicals using modified cyclodextrins, *J. High Resolut. Chromatogr.* 14 (1991) 530–536.

260. W. A. König, D. Icheln, and I. Hardt, Unusual retention behaviour of methyl lactate and methyl 2-hydroxybutyrate enantiomers on a modified cyclodextrin, *J. High Resolut. Chromatogr.* 14 (1991) 694–695.

261. W. Melander, D. E. Campbell, and C. Horváth, Enthalpy-entropy compensation in reversed-phase chromatography, *J. Chromatogr.* 158 (1978) 215–225.

262. M. Schneider and K. Ballschmiter, Alkyl nitrates as achiral and chiral solute probes in gas chromatography: novel properties of a β-cyclodextrin derivative and characterization of its enantioselective forces, *J. Chromatogr. A* 852 (1999) 525–534.

263. C. Bicchi, E. Liberto, C. Cagliero, C. Cordero, B. Sgorbini, and P. Rubiolo, Conventional and narrow bore short capillary columns with cyclodextrin derivatives as chiral selectors to speed-up enantioselective gas chromatography and enantioselective gas chromatography-mass spectrometry analyses, *J. Chromatogr. A* 1212 (2008) 114–123.

264. A. C. Evans, C. Meinert, J. H. Bredehöft, C. Giri, N. C. Jones, S. V. Hoffmann, and U. J. Meierhenrich, Anisotropy spectra for enantiomeric differentiation of biomolecular building blocks, *Top. Curr. Chem.* 341 (2013) 271–300.

5 Analysis of Dynamic Phenomena in Liquid Chromatographic Systems with Reactions in the Mobile Phase

Lei Ling and Nien-Hwa Linda Wang

5.1   INTRODUCTION

5.1.1   Reactions and Separations in Liquid Chromatography

In many liquid chromatographic systems, reactions can occur among the solutes in the mobile phase, resulting in dynamic phenomena that are different from those in nonreactive systems. Some reactions are desirable and have important applications. For instance, one can analyze the peak areas of the reactants and the products to estimate the rate constants for irreversible reactions [1,2] or the reaction stoichiometry and the equilibrium constants for reversible reactions [3]. For a reversible reaction from a reactant to one or two products, if the products are well separated from the reactant and from each other, the conversion can be higher than the equilibrium conversion in a batch reactor [4–8]. For the separation of enantiomers, if the sorbent does not have sufficient enantioselectivity, one can increase peak resolution by adding a ligand to the mobile phase [9–12]. The ligand can form complexes with the enantiomers, which adsorb differently to allow separation. For the separation of lanthanide ions, a chelating agent (complexant) can be added in the mobile phase to separate lanthanides with identical valence and a small size difference [13]. The complexant can form complexes with lanthanides with largely different equilibrium constants (or high selectivity). The chelating agent increases the partition of the preferred lanthanides in the mobile phase and thus increases their migration speeds and enables their separation from the less-preferred lanthanides.

Some other reactions are undesirable, however, in a liquid chromatographic system. The isomerization of peptides and the aggregation of proteins and polymers give rise to split or merged peaks in elution chromatography under different conditions [3,14,15]. The number of peaks, the relative peak areas, and the peak retention times can vary with temperature, flow rate, column length, sample concentration, and sample volume. The complex elution behavior greatly increases the difficulties in sample analysis. In capture chromatography, protein aggregation may cause an early breakthrough and thus a lower dynamic binding capacity [16]. Protein denaturation can occur when contacting with reversed-phase or hydrophobic sorbents [17], leading to multiple peaks in analytical chromatography or a lower recovery in preparative chromatography.

5.1.2   Objectives

In this chapter, the dynamic phenomena of liquid chromatographic systems with six typical reactions in the mobile phase are analyzed. The first goal is to understand the dynamic adsorption and separation mechanisms in such systems. The second goal is to identify the key factors that control the conversion and resolution of the reacting species in elution and frontal chromatography. The third goal is to provide overall guidelines for design and optimization of reactive chromatographic systems. In the examples, we use data from literature sources and report several novel results from our own calculations.

5.1.3   Methods and Strategies

To achieve the goals of this study, we first identified all the independent dimensionless groups from the rate model equations, which included differential mass balance equations, boundary conditions, and initial conditions. We then tested the rate model and parameters by comparing simulated chromatograms with the experimental data for the six cases reported in the literature. Once the simulation results agreed closely with the data, the models and parameters were used to simulate chromatograms over a wide range of dimensionless group values. The results were used to identify the key dimensionless groups that controlled the conversion, the yield, and the splitting or merging of the peaks or fronts of the reactive species. Various reaction and separation phenomena were analyzed and classified based on the key dimensionless groups. Dynamic column profiles were used to better understand the mechanisms of reactions and separations in typical systems.

5.1.4 Overview of the Types of Reactions Considered

Six typical cases with reactions occurring in the column were analyzed and are discussed in detail in Sections 5.3–5.9 (Table 5.1 and Figure 5.1). In Case I, we discuss how to use elution chromatography to measure the rate constant for a first-order irreversible reaction. We identify the key dimensionless groups that control the conversion of the reactants and the resolution of the peaks. In Case II, we identify the factors that control peak-splitting or peak-merging phenomena for a first-order reversible reaction. We introduce a way to estimate the equilibrium constant. In Case III, we show how to force a reversible decomposition reaction to reach complete conversion and how to obtain pure products at the same time. In Cases IV and V, we show how protein aggregation can cause split or merged peaks in elution chromatography and multiple plateaus in frontal chromatography. We introduce a method for estimating the reaction stoichiometry and equilibrium constants. In Case VI, we identify the sorbent and the ligand/complexant properties that are important for separation.

TABLE 5.1
Types of Reactions and Isotherms in Six Cases

Case	Reaction	Isotherm	Type	Example	Reference
I	$A \rightarrow B$	Langmuir (linear)	Pulse	Initial step of tetrachloroterephthaloyl chloride esterification	1
II	$A \leftrightarrow B$	Langmuir (linear)	Pulse	Peptide isomerization	14
III	$A \leftrightarrow B + C$	Freundlich (nonlinear)	Pulse	Methyl formate hydrolysis	4
IV(A)	$2M \leftrightarrow D$	Size exclusion	Pulse	Protein dimerization in SEC	28
IV(B)	$2M \leftrightarrow D$	Langmuir (nonlinear)	Frontal	Protein dimerization in IMAC	16
V	$2T \leftrightarrow O$ $T + O \leftrightarrow D$	Langmuir (nonlinear)	Pulse	Protein multiaggregation in HIC	3,15
VI	$L + R \leftrightarrow LR$ $L + S \leftrightarrow LS$	Langmuir (linear)	Pulse	Mobile phase complexation for chiral separation in UPLC	12

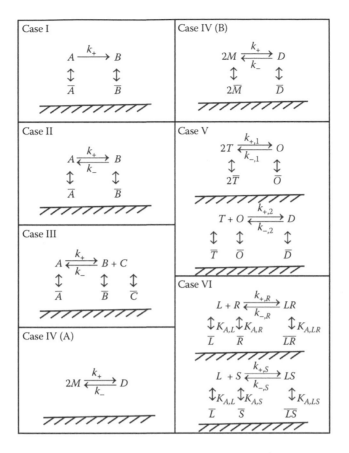

FIGURE 5.1 Overview of six mobile phase reactions in chromatography.

This chapter aims to answer the following key questions: (1) What are the differences in chromatographic behavior between reactive and nonreactive systems? (2) How can reaction stoichiometry, equilibrium constants, and rate constants be determined? (3) For desirable reactions, how can one improve conversion and product purity? (4) For undesirable reactions, how can one reduce the adverse effect of the reactions? (5) For ligand- or complexant-assisted separations, how can a ligand and a sorbent work synergistically to separate solutes with small differences in structures or properties?

5.2 VERSATILE REACTION–SEPARATION MODEL AND SIMULATIONS

5.2.1 OVERVIEW

The effluent histories and dynamic concentration profiles as a function of time can be useful for fundamental understanding of various complex phenomena in a chromatographic system with some reactions taking place. The propagation of concentration waves in a nonlinear adsorption system with significant mass transfer effects is affected by competitive adsorption in the sorbent phase, convection and dispersion in the mobile phase, and any reactions occurring in the mobile phase or in the adsorbent phase. The detailed dynamics for systems with nonlinear isotherms cannot be obtained from analytical solutions. They can only be obtained from the numerical solutions of rate model equations, which consist of the differential mass balance equations for all the components, the competitive adsorption rate equations or the equilibrium adsorption isotherm equations, the initial conditions, and the boundary conditions, which are related to the operating conditions of the various types of chromatography. The rate model VERSE (Versatile Reaction–Separation) and related simulations based on pore diffusion were first developed for batch chromatography by Wang and associates in 1991 [18].

VERSE is an expanded version of an earlier rate model for batch chromatography without reactions, which took into account axial dispersion, film mass transfer, intraparticle pore diffusion, and equilibrium competitive adsorption and ion exchange [19]. To model complex protein adsorption and desorption phenomena in chromatography, the rate model was modified to include nonequilibrium (or slow) adsorption and desorption [20], aggregation reactions in the mobile phase [15,16], and denaturation reactions in the stationary phase [17].

The pore diffusion model, however, was found to be inaccurate for high-affinity systems at high loading, where surface diffusion effects dominated wave spreading. Surface diffusion can result in asymmetric breakthrough curves with a sharp rise at low concentrations and then a slow approach to saturation at high concentrations. As adsorbed phase solute concentration approaches saturation, the driving force for surface diffusion diminishes, resulting in the slow approach to saturation. In contrast, the driving force for pore diffusion does not have this limitation, resulting in symmetric breakthrough curves, except for systems in which the diffusion rate is much smaller than the convection rate. Ma et al. [21] revised the model equations to include surface diffusion and parallel pore and surface diffusion so that the model can describe accurately the breakthrough curves at high loading. At a low loading, both the pore diffusion model and the surface diffusion model give closely similar wave spreading. Since this chapter focuses on the reaction phenomena, the pore diffusion model is used.

Koh et al. [22,23] expanded the parallel pore and surface diffusion model in VERSE to fluidized and expanded beds. Ernest et al. [24,25] further expanded the VERSE model for single-column batch chromatography to carousel systems. Hritzko et al. [26] expanded and experimentally verified the VERSE model for simulated moving-bed (SMB) systems. The VERSE models have been tested and found to be in close agreement with the experimental data in more than 30 systems, including the separations of ions and small organic compounds, amino acids, lactic acid, sugars, chiral compounds, antibiotics, paclitaxel, insulin, and other proteins [27].

In summary, the general VERSE model and simulations take into account various types of adsorption mechanisms (adsorption, ion exchange, and size exclusion); detailed mass transfer effects (convection, axial dispersion, film mass transfer, intraparticle pore diffusion, surface diffusion, or parallel pore and surface diffusion); slow adsorption and desorption; and any chemical reactions in both the solution phase and the solid phase in chromatography (Figure 5.2). Both equilibrium

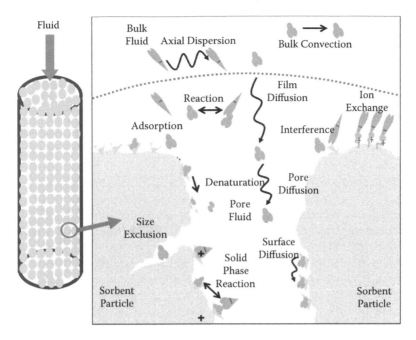

FIGURE 5.2 (See Color Insert.)   Various reactions and mass transfer mechanisms considered in VERSE.

and nonequilibrium adsorption can be considered. Reactions include dimerization, trimerization, mixed aggregation, and denaturation in the solution phase or on the surface of the stationary phase. It can simulate chromatographic processes in fixed beds, expanded beds, fluidized beds, carousel, and SMB. VERSE also allows for frontal chromatography, isocratic or gradient elution chromatography, elution with periodic total recycle, displacement chromatography, and other cyclic operations that involve multiple concentration changes in a column inlet as a function of time.

In the simulations reported in this chapter, a simplified VERSE model and simulations based on local adsorption equilibrium, pore diffusion, and reactions in the mobile phase for batch chromatography were used. The key equations are introduced in Section 5.2.2. The models for different types of reaction, isotherm, and mass transfer effects are introduced in Section 5.2.3. The significance of dimensionless groups is discussed in Section 5.2.4. The chromatographic operating modes that can be simulated in VERSE are introduced in Section 5.2.5. The numerical methods applied in VERSE simulations are briefly discussed in Section 5.2.6.

5.2.2 THE RATE MODEL EQUATIONS FOR SIMULATIONS IN THIS CHAPTER

The key assumptions in the rate model are the following:

1. The column is packed uniformly with spherical particles of the same size.
2. The flow distribution in the column is uniform.

3. The column is isothermal.
4. The concentration gradients in the bulk phase occur only in the axial direction.
5. The concentration gradients inside the particles occur only in the radial direction.
6. The diffusivities are constant and independent of concentrations.
7. Reactions only occur in the mobile phase.
8. The intrinsic adsorption and desorption rates are much faster than the mass transfer rates, such that the local concentrations of the adsorbed solutes are related to the local concentrations of the solutes in the pore phase by adsorption equilibrium isotherms.
9. Pore diffusion is the dominant intraparticle diffusion mechanism.

The general differential mass balance equations for each solute in the bulk phase, the pore phase, and the solid phase for the pore diffusion model were derived previously [18]. These equations take into account the time-dependent generation or consumption of any species by reactions in the mobile phase. In the numerical solution, the variables in the equations are in dimensional form, and the equations are solved with the proper boundary and initial conditions. The concentrations of a species in the bulk phase $C_{b,i}$, in the pore phase $C_{p,i}$, and in the solid phase $\overline{C_{p,i}}$ are determined as a function of time t, the axial distance from the column inlet z, and the radial position within a particle r. The solutions give the effluent histories and the concentration profiles in a column as a function of time.

The equations of the pore diffusion model are described in the following in terms of dimensionless variables to facilitate the discussion of dimensionless groups in Section 5.2.4. The dimensionless variables are

$$x \equiv \frac{z}{L} \tag{5.1a}$$

$$\xi \equiv \frac{r}{R} \tag{5.1b}$$

$$\theta \equiv \frac{t}{\tau} \tag{5.1c}$$

$$\tau \equiv \frac{L}{u_0} = \frac{L\varepsilon_b}{u_s} \tag{5.1d}$$

$$c_{b,i} \equiv \frac{C_{b,i}}{C_{f,i}} \tag{5.2a}$$

$$c_{p,i} \equiv \frac{C_{p,i}}{C_{f,i}} \tag{5.2b}$$

$$\overline{c_{p,i}} \equiv \frac{\overline{C_{p,i}}}{C_{T,i}} \tag{5.2c}$$

$$\varphi_{L,i} \equiv \frac{C_{f,i}}{C_{T,i}(1-\varepsilon_p)} \tag{5.2d}$$

where L is the column length, R is the sorbent particle radius, ε_b is the interparticle void fraction, ε_p is the intraparticle void fraction (porosity), u_s is the mobile phase superficial velocity, and u_0 is the mobile phase interstitial velocity ($u_0 = u_s/\varepsilon_b$). For a nonreactive system, $C_{f,i}$ is the maximum feed concentration of component i; for systems with reactions, $C_{f,i}$ is the maximum feed concentration of the major reactant. $C_{T,i}$ is the maximum capacity for component i.

The differential mass balance equations with boundary and initial conditions for the dimensionless bulk phase concentrations of species i are

$$\frac{\partial c_{b,i}}{\partial \theta} = \frac{1}{Pe_{b,i}} \frac{\partial^2 c_{b,i}}{\partial x^2} - \frac{\partial c_{b,i}}{\partial x} + Y_{b,i} - N_{f,i}(c_{b,i} - c_{p,i,\xi=1}) \tag{5.3a}$$

$$x = 0, \quad \frac{\partial c_{b,i}}{\partial x} = Pe_{b,i}(c_{b,i} - c_{f,i}(\theta)) \tag{5.3b}$$

$$x = 1, \quad \frac{\partial c_{b,i}}{\partial x} = 0 \tag{5.3c}$$

$$\theta = 0, \quad c_{b,i} = c_{b,i}(0, x) \tag{5.3d}$$

For the concentrations of species i in the pore phase, the respective equations are

$$Ke_i\left[\varepsilon_p \frac{\partial c_{p,i}}{\partial \theta} - \varepsilon_p Y_{p,i}\right] = N_{p,i} \frac{1}{\xi^2} \frac{\partial}{\partial \xi}(\xi^2 \frac{\partial c_{p,i}}{\partial \xi}) - \frac{Y_{l,i}}{\varphi_{L,i}} \tag{5.4a}$$

$$\xi = 0, \quad \frac{\partial c_{p,i}}{\partial \xi} = 0 \tag{5.4b}$$

$$\xi = 1, \quad \frac{\partial c_{p,i}}{\partial \xi} = Bi_i(c_{b,i} - c_{p,i}) \tag{5.4c}$$

$$\theta = 0, \quad c_{p,i} = c_{p,i}(0, \xi) \tag{5.4d}$$

The term Ke_i is the fraction of the pore volume that is accessible by component i. $Y_{b,i}$, $Y_{p,i}$ represent the generation of component i by reaction in the bulk phase and in the pore phase, respectively. $Y_{l,i}$ describes the net loss of component i in the pore phase by adsorption onto the solid phase. Under the local equilibrium condition, the solid phase concentrations are related to the pore phase concentrations via the adsorption isotherms. $Y_{l,i}$ in Equation (5.4a) can be calculated from the pore phase concentration and the adsorption isotherms in the following equation:

$$Y_{l,i} = \sum_{j=1}^{N}\left[\frac{\overline{\partial c_{p,i}}}{\partial c_{p,j}} \frac{\partial c_{p,j}}{\partial \theta}\right] \tag{5.5}$$

The mixing in the extracolumn dead volume is modeled by a completely stirred tank reactor (CSTR) attached at each column inlet and outlet.

$$V_D \frac{dC_{out,i}}{dt} = F\left(C_{in,i} - C_{out,i}\right) \tag{5.6}$$

where V_D is the extracolumn dead volume, F is the flow rate, and $C_{in,i}$ and $C_{out,i}$ are the concentrations of species i at the CSTR inlet and outlet, respectively. In the simulations, the CSTR dead volumes were set to be less than 1% of the column volume. The dimensionless groups Pe_b, N_f, N_p, and Bi are defined in a table and are in Section 5.2.4.

5.2.3 Models for Reaction, Isotherm, and Mass Transfer

To efficiently solve a large number of mass balance equations, a FORTRAN program was written for the general VERSE model. In this program, a wide range of reaction, isotherm, and mass transfer models are available in separate modules for users to choose.

5.2.3.1 Reaction Models Considered

A major advantage of the VERSE model over other rate models is that it can take into account reactions in chromatography. Various elementary reaction models, such as dimerization [28], trimerization [29], denaturation [17], and heteroaggregation [16], have been incorporated into the VERSE program. In addition, other more complex sequential and parallel reactions can be modeled as combinations of these elementary reactions.

A general expression for the generation term with m reactions in the bulk phase is

$$Y_{b,i} = \frac{L}{u_0} \sum_m \frac{\sigma_{i,m}}{C_{f,i}} \left(k_{+m} \prod_j C_{b,j}^{-\sigma_{i,m}} - k_{-m} \prod_j C_{b,j}^{'-\sigma_{i,m}} \right) \tag{5.7}$$

where σ_i is the coefficient of component i in a reaction (positive for products and negative for reactants). Each reaction is generally assumed to be reversible and follow the reaction stoichiometry. The k_{+m} is the forward rate constant, and k_{-m} is the reverse rate constant of the mth reaction. $C_{b,j}$ and $C_{b,j}'$ are the concentrations of reactants and products. Similar sets of equations can be written for pore phase reactions or for solid phase reactions.

In this chapter, six cases with various types of reactions in the bulk phase and in the pore phase are presented. The detailed expressions of Y_b or Y_p for each case are listed in Table 5.2. No examples of solid phase reactions are shown as they are discussed elsewhere [17].

5.2.3.2 Isotherm Models in VERSE

When the adsorption and desorption rates are much faster than the convection and mass transfer rates, there is local equilibrium between the solid phase concentrations and the pore phase concentrations. Various equilibrium isotherm models for different

TABLE 5.2

Reaction Rate Models and Dimensionless Groups for Six Reactions in the Bulk and Pore Phases

Case	Reaction	$Y_{b,i}$ or $Y_{p,i}$	N_{k+}	N_{k-}
I	$A \to B$	$Y_A = -\dfrac{Lk_+C_A}{u_0C_{f,A}}, Y_B = -Y_A$	$\dfrac{Lk_+}{u_0}$	N.A.
II	$A \leftrightarrow B$	$Y_A = -\dfrac{L(k_+C_A - k_-C_B)}{u_0C_{f,A}}, Y_B = -Y_A$	$\dfrac{Lk_+}{u_0}$	$\dfrac{Lk_-}{u_0}$
III	$A \leftrightarrow B + C$	$Y_A = -\dfrac{L(k_+C_A - k_-C_BC_C)}{u_0C_{f,A}}, Y_B = Y_C = -Y_A$	$\dfrac{Lk_+}{u_0}$	$\dfrac{Lk_-C_{f,A}}{u_0}$
IV	$2M \leftrightarrow D$	$Y_M = -\dfrac{2L(k_+C_M^2 - k_-C_D)}{u_0C_{f,M}}, Y_D = -\dfrac{Y_M}{2}$	$\dfrac{Lk_+C_{f,M}}{u_0}$	$\dfrac{Lk_-}{u_0}$
V	$2T \leftrightarrow O$ (1) $T + O \leftrightarrow D$ (2)	$Y_T = -\dfrac{L(2k_{+,1}C_T^2 - 2k_{-,1}C_O + k_{+,2}C_TC_O - k_{-,2}C_D)}{u_0C_{f,T}}$ $Y_O = \dfrac{L(k_{+,1}C_T^2 - k_{-,1}C_O - k_{+,2}C_TC_O + k_{-,2}C_D)}{u_0C_{f,T}}$ $Y_D = \dfrac{L(k_{+,2}C_TC_O - k_{-,2}C_D)}{u_0C_{f,T}}$	$\dfrac{Lk_+C_{f,T}}{u_0}$	$\dfrac{Lk_-}{u_0}$
VI	$L + R \leftrightarrow LR$ (1) $L + S \leftrightarrow LS$ (2)	$Y_L = -\dfrac{L(k_{+,R}C_LC_R + k_{+,S}C_LC_S - k_{-,R}C_{LR} - k_{-,S}C_{LS})}{u_0C_{f,L}}$ $Y_R = -\dfrac{L(k_{+,R}C_LC_R - k_{-,R}C_{LR})}{u_0C_{f,L}}, Y_{LR} = -Y_R$ $Y_S = -\dfrac{L(k_{+,S}C_LC_S - k_{-,S}C_{LS})}{u_0C_{f,L}}, Y_{LS} = -Y_S$	$\dfrac{Lk_+C_{f,L}}{u_0}$	$\dfrac{Lk_-}{u_0}$

adsorption mechanisms are available in VERSE. The isotherms include the constant separation factor model [30], the reverse-phase modulator model [31], the multicomponent Langmuir model [32–34], the Bi-Langmuir model [35], the Freundlich isotherm model [24,25], the multicomponent mass action model [24,25], and others. The most widely used model is the multicomponent Langmuir isotherm model, which was used in the simulations presented in this chapter unless noted otherwise:

$$\overline{C_{p,i}} = \frac{a_iC_{p,i}}{1+\displaystyle\sum_{j=1}^{N}b_jC_{p,j}} \tag{5.8}$$

When the adsorption and desorption rates are relatively slow compared to the mass transfer rates, then one uses the Langmuir–Hinshelwood equation, which reduces to Equation (5.8) at equilibrium [20].

5.2.3.3 Mass Transfer Models

Three types of mass transfer effects are considered in the governing equations: axial dispersion, film mass transfer, and intraparticle diffusion. The last can be pore diffusion, surface diffusion, or both in parallel. Previous studies showed that in the linear isotherm region, the effects of surface diffusion on frontal or elution chromatography cannot be distinguished from those of pore diffusion [21,31]. In this chapter, we consider pore diffusion as the only intraparticle diffusion mechanism.

The axial dispersion coefficient E_b can be determined experimentally or estimated from literature correlations. One can calculate the Peclet number $Pe_b = u_0 L/E_b$, which represents the ratio of the convection rate to the axial dispersion rate. The Chung and Wen correlation [36] is used to estimate the axial dispersion coefficient E_b in the simulations

$$Pe_b = \frac{L}{2\varepsilon_b R}(0.2 + 0.011 Re^{0.48})$$ (5.9a)

$$Re \equiv \frac{2R\rho u_0 \varepsilon_b}{\mu}$$ (5.9b)

where ρ and μ are the density and the viscosity of the fluid, respectively. When Re is less than 1, Pe_b is reduced to $0.1\ L/R\varepsilon_b$, according to Equation (5.9a).

The film mass transfer coefficient k_f is a function of the Reynolds number Re and the Schmidt number Sc. It can be estimated from literature correlations. Since Re is generally less than 0.1 in the systems of interest, the correlation of Wilson and Geankoplis [37], which is valid for low Re values (<0.1), is used:

$$J \equiv \left[\frac{k_{f,i}}{u_0 \varepsilon_b} \right] Sc_i^{\frac{2}{3}} = \frac{1.09}{\varepsilon_b} Re^{-\frac{2}{3}}$$ (5.10a)

$$Sc_i \equiv \frac{\mu}{\rho D_{b,i}}$$ (5.10b)

where $D_{b,i}$ is the Brownian diffusivity of component i.

The intraparticle pore diffusivity D_p can be estimated by fitting pulse data for low-affinity solutes or frontal data for high-affinity solutes to the simulation results. A rough initial guess of $D_{p,i}$ can be obtained from the correlation of Mackie and Meares [38]:

$$D_{p,i} = \left[\frac{\varepsilon_p}{2 - \varepsilon_p} \right]^2 \frac{D_{b,i}}{\varepsilon_p}$$ (5.11)

5.2.4 KEY DIMENSIONLESS GROUPS

Dimensionless groups can be derived from the coefficients of the dimensionless differential mass balance equations and the initial and boundary conditions. Each coefficient is a combination of different physical variables, such as velocity, particle size, column length, solute concentration, and reaction rate constants. Values of all the coefficients affect the solutions of the rate model equations. As long as the coefficients in the equations, or the dimensionless groups, are fixed, the solutions of the equations with a fixed set of initial and boundary conditions are uniquely determined, although the values of the individual physical variables are different. Therefore, if we show the effects of the dimensionless groups, rather than the individual variables, on reaction and separation, we can greatly reduce the number of parameters to be analyzed.

A characteristic convection rate, which can be easily measured and controlled, is chosen to be the benchmark for the dimensionless rates. The reaction rates relative to the convection rate (N_{k+}, N_{k-}) for the six cases are shown in Table 5.2. The mass transfer rates relative to the convection rate are shown in Table 5.3. These dimensionless rates, as shown in Sections 5.3–5.9, control the chromatographic behavior.

The inverse of the Peclet number $1/Pe_b$ is the axial dispersion rate relative to the convection rate, N_f is the film mass transfer rate relative to the convection rate, N_p is the intraparticle diffusion rate relative to the convection rate. Alternatively, one can combine the phase ratio $P = (1 - \varepsilon_b)/\varepsilon_b$ with N_p to form a new dimensionless group, $N_D (= PN_p)$. Since intraparticle diffusion occurs in adsorbent particles and convection occurs in interstitial void, the product PN_p takes into account any difference in ε_b. For this reason, we use N_D in the analysis to represent the intraparticle diffusion rate relative to the convection rate. Bi, which appears in the boundary condition of pore phase mass balance equations, is the film mass transfer rate relative to the intraparticle diffusion rate, as shown in Table 5.3. Bi is related to N_f and N_D ($Bi = N_f/3N_D$). Therefore, it is not an independent dimensionless group. Either Bi or N_f can be used to indicate the importance of film diffusion.

If the adsorption isotherm is in the linear region, the effects of axial dispersion, film mass transfer, intraparticle diffusion, and slow adsorption and desorption on peak spreading can be estimated from the plate height equation [39]. For a local equilibrium system, a simplified equation for HETP is

$$\frac{HETP}{L} = \frac{2}{Pe_b} + \left[\frac{(1-\varepsilon_b)\varepsilon_p + a}{(1-\varepsilon_b)\varepsilon_p + a + \varepsilon_b} \right]^2 \left(\frac{2}{15N_D} + \frac{2}{N_f} \right) \tag{5.12}$$

TABLE 5.3

Dimensionless Groups of Mass Transfer Rates

Axial Dispersion Rate / *Convection Rate*	$\dfrac{1}{Pe_{b,i}} = \dfrac{E_{b,i}}{u_0 L}$
Film Mass Transfer Rate / *Convection Rate*	$N_{f,i} = 3\left(\dfrac{L}{R}\right)\dfrac{(1-\varepsilon_b)k_{f,i}}{\varepsilon_b u_0}$
Intra – particle Diffusion Rate / *Convection Rate*	$N_{p,i} = \dfrac{L\varepsilon_p D_{p,i}}{u_0 R^2}, N_{D,i} = \dfrac{(1-\varepsilon_b)\varepsilon_p L D_{p,i}}{\varepsilon_b u_0 R^2}$
Film Mass Transfer Rate / *Intra – particle Diffusion Rate*	$Bi_i = \dfrac{k_{f,i} R}{D_{p,i}\varepsilon_p} = \dfrac{N_{f,i}}{3N_{D,i}}$

---

where HETP is the theoretical plate height. For an adsorptive system with a sufficiently large "a" value, the term in the bracket is close to 1. This equation allows one to evaluate the relative importance of axial dispersion, intraparticle diffusion, and film diffusion from the values of Pe_b, $15N_D$, and N_f.

The loading factor, L_f for an adsorbing system, is the ratio of the loading amount to the sorbent capacity at a given feed concentration C_f:

$$L_f \equiv \frac{C_f V_f}{q_f V_C} = \frac{C_f u_s t_L}{q_f L}$$

(5.13)

where q_f is the solid phase concentration in equilibrium with C_f, V_f is the feed volume, V_C is the column packing volume, and t_L is the loading time. For a non-adsorbing system, L_f is the ratio of the loading volume to the column packing volume V_C:

$$L_f \equiv \frac{V_f}{V_C} = \frac{u_s t_L}{L}$$

(5.14)

In a pulse test, if the column length or the velocity changes, the same chromatogram can be obtained only if the loading time is changed accordingly to keep the values of L_f and all the other dimensionless groups constant.

5.2.5 MODES OF CHROMATOGRAPHY

There are several modes of batch chromatography with different initial and boundary conditions. Frontal chromatography involves a step change in the concentration at the column inlet at time $t = 0$. Isocratic elution chromatography involves two step changes in the inlet feed concentration at $t = 0$ and at $t = t_L$. In gradient elution or stepwise elution, the inlet mobile phase composition after the pulse injection is a function of time. In displacement chromatography, the column is preequilibrated with a weakly adsorbing solute. After feed loading, the inlet condition involves a step increase in the concentration of a high-affinity displacer in the mobile phase.

Different chromatographic modes have been simulated using VERSE: frontal [15,31], isocratic elution [16], stepwise or linear gradient elution [31], displacement [16], and other cyclic operations involving multiple changes in the flow rate or the concentrations at the column inlet as a function of time. For multicolumn runs, one can also simulate carousel [24–26] and SMB processes [32–35,40–46]. In this chapter, we focus on simulations of frontal or isocratic elution in batch chromatography.

5.2.6 NUMERICAL METHODS

The differential mass balance equations in the VERSE model (Section 5.2.2) are discretized using orthogonal collocation on fixed finite elements [47,48]. The normalized space coordinates in the axial and particle radial directions are divided into discrete elements, within which a number of interior collocation points or nodes are chosen. In the axial direction, Legendre polynomials are used to discretize and approximate the concentration profile. In the particle radial direction, only one element is used, and Jacobi polynomials are employed because they satisfy the boundary condition at the interior node of the element automatically. Thus, the partial differential mass balance equations are converted to a set of ordinary differential equations.

In the bulk phase, Equations (5.3a)–(5.3d) become

$$\frac{\partial c_{b,i,h}}{\partial \theta} = \frac{1}{Pe_{b,i}} \sum_{k=1}^{n_a+2} B_{Lh,k} c_{b,i,k} - \sum_{k=1}^{n_a+2} A_{Lh,k} c_{b,i,k} + Y_{b,i,k} - N_{f,i}(c_{b,i,h} - c_{p,i,h,\xi=1}) \tag{5.15a}$$

$$x = 0, \quad \sum_{k=1}^{n_a+2} A_{Lh,k} c_{b,i,k} = Pe_{b,i}(c_{b,i,h} - c_{f,i}(\theta)) \tag{5.15b}$$

$$x = 1, \quad \sum_{k=1}^{n_a+2} A_{Lh,k} c_{b,i,k} = 0 \tag{5.15c}$$

$$\theta = 0, \quad c_{b,i,h} = c_{b,i,h}(0, x) \tag{5.15d}$$

In the pore phase, Equations (5.4a)–(5.4d) become

$$Ke_i \left[\varepsilon_p \frac{\partial c_{p,i,h}}{\partial \theta} - \varepsilon_p Y_{p,i,h} \right] = N_{p,i} \sum_{k=1}^{n_p+1} B_{Jh,k} c_{p,i,k} - \frac{Y_{l,i,h}}{\varphi_{L,i}} \tag{5.16a}$$

$$\xi = 0, \quad \sum_{k=1}^{n_p+1} A_{Jh,k} c_{p,i,k} = 0 \tag{5.16b}$$

$$\xi = 1, \quad \sum_{k=1}^{n_p+1} A_{Jh,k} c_{p,i,k} = Bi_i(c_{b,i,h} - c_{p,i,h}) \tag{5.16c}$$

$$\theta = 0, \quad c_{p,i,h} = c_{p,i,h}(0, \xi) \tag{5.16d}$$

where A and B are the discretization matrices approximating the first and the second derivatives, respectively. The subscript L represents Legendre polynomials, and J represents Jacobi polynomials, n_a is the number of interior axial collocation points; n_p is the number of interior particle collocation points; and h represents a specific collocation point within either an axial element or within a particle.

The system is solved with a time integrator, DASSL, which is a well-known differential/algebraic system solver developed by Petzold [49]. The solution scheme is robust even for complex sets of coupled stiff equations.

A new user-friendly interface has been developed for the VERSE program. It allows a user to select the isotherm, the reaction, and the mass transfer model of interest from pull-down menus. The parameter values are entered using the interface. The following input parameters are required for a numerical solution or simulation of batch chromatography: (a) the intrinsic parameters, which include particle size, bed and particle void fractions, adsorption isotherms or rate equations, Brownian diffusivities, intraparticle pore or surface diffusivities, and reaction parameters; (b) the system parameters, which include column diameter, column length, and dead volume; (c) the initial column concentration profiles; (d) the flow rate and composition of the inlet stream as a function of time; and (e) the numerical simulation parameters, which include the number of axial elements, the number of collocation points in each axial element, the number of particle collocation points, the maximum time step for integration, the relative tolerance, and the absolute tolerance for each component. One should also specify desired outputs, such as the bulk phase or pore phase concentration profiles at certain times, and the effluent histories at certain positions. The product purity and yield can be readily calculated from the simulated effluent histories, which are displayed along with column profiles in the interface.

5.3   CASE I: FIRST-ORDER IRREVERSIBLE REACTION

5.3.1   SYSTEM CONSIDERED AND MODEL TESTING

The kinetic rate constants for the esterification of tetrachloroterephthaloyl chloride in a C18 reversed-phase column were estimated [1,50,51]. The overall reaction was catalyzed by pyridine or 4-picoline. The mechanism included four irreversible reactions [52]. The first reaction was found to be first order and much faster than the other three. Since the elution time was chosen to be comparable to the reaction time of the first reaction, the other reactions were assumed to have no impact. The effective rate constant k_{app} was determined from chromatographic data using the "inert standard method." The relevant equations are

$$\ln\left[\frac{A_R}{A_I}\right] = \ln\left[\frac{A_R}{A_I}\right]_{t=0} - k_{app}t_R \tag{5.17a}$$

$$k_{app} = \left(\frac{t_m}{t_R}\right)k_m + \left(\frac{t_s}{t_R}\right)k_s \tag{5.17b}$$

where A_R and A_I are the peak areas of the reactant and the inert material, respectively; t_R is the total retention time; t_m and t_s are the retention times in the mobile phase and the solid phase, respectively; and k_m and k_s are the rate constants in the mobile phase and the solid phase, respectively. To simplify the simulations, any solid phase reaction was ignored, as depicted in Figure 5.1. Then, $k_s = 0$.

The effective value of k_m at 25°C is 0.042 min^{-1} (Table 5.4). The adsorption isotherm was assumed to be linear. The retention factors were calculated from the retention volume. The parameters required for the VERSE simulations are listed in Table 5.4. Two simulated chromatograms at different flow rates are shown in Figures 5.3a and 5.3b. They agree fairly well with the experimental data. The thus-validated models and parameters are used in the simulations that follow.

5.3.2 Improvement of Conversion and Resolution

As shown in Figure 5.3c, when the physical variables (k_+, F, L, R, and V) vary but the dimensionless groups are kept the same, the simulated chromatogram remains the same as in Figure 5.3a. The result indicates that the conversion and the resolution are fully determined by the dimensionless groups and not by the specific values of the dimensional physical variables.

Bolme and Langer [1] showed that as the flow rate decreases, the residence time of the reactant in a given column increases, and the conversion increases. Detailed simulations showed that as the reaction rate relative to the convection rate increases,

---

TABLE 5.4

Simulation Parameters and Dimensionless Groups for Case I

System Parameters

L (cm)	ID (cm)	R (μm)	ε_b	ε_p
50	0.46	2.5	0.35	0.28

Operating Conditions

Figure 5.3	F (mL/min)	$C_{i,A}$ (mM)	t_L (min)
(a)	0.23	0.6	0.087
(b)	0.9	0.6	0.022

Reaction Parameters

Reaction	k_+ (min^{-1})	k_- (min^{-1})
$A \rightarrow B$	0.042	0

Isotherm Parameters (Langmuir)

Component	a	b (mM^{-1})
A	0.49	0
B	0.29	0

Mass Transfer Parameters

Component	D_b (cm^2/min)	D_p (cm^2/min)	k_f (cm/min)	E_b (cm^2/min)
A and B	0.001	0.0005	From [37]	From [36]

Numerical Parameters

		Collocation Points		Tolerance	
Axial Elements	Step Size (L/u_0)	Axial	Particle	Absolute (mM)	Relative
200	0.001	4	2	10^{-4}	10^{-4}

Dimensionless Groups

Figure 5.3	N_{k+}	N_D	Pe_b	N_f	L_f
(a)	0.531	5.3×10^4	5.7×10^4	5.7×10^4	0.0049
(b)	0.136	1.3×10^4	5.7×10^4	6.3×10^5	0.0049

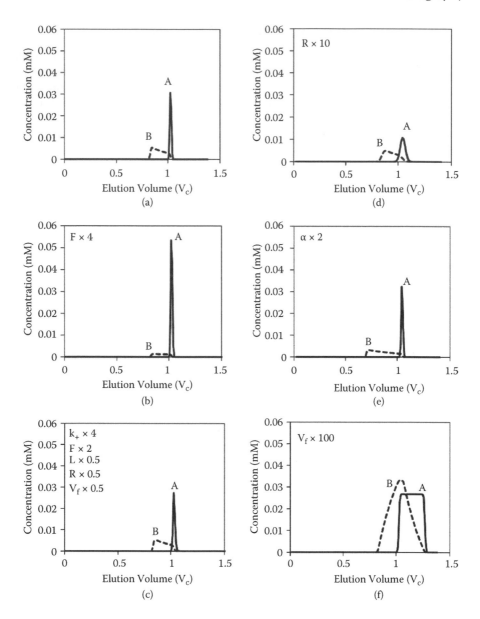

FIGURE 5.3 Simulated chromatograms for Case I. A is the reactant, and B is the product; (a) and (b) are for model testing. The parameters used in the simulations and the four dimensionless groups are given in Table 5.4. (c) has different physical variables but the same values of the dimensionless groups as (a). (d), (e), and (f) have different particle size, selectivity, and loading amount, respectively, compared to (a). The values of dimensionless groups for (d), (e), and (f) are $N_{k+} = 0.531$, $N_D = 5.3 \times 10^4$, $Pe_b = 5.7 \times 10^4$, $N_f = 1.6 \times 10^6$, $L_f = 0.0049$, except in (d) $N_D = 5.3 \times 10^2$, $Pe_b = 5.7 \times 10^3$, $N_f = 3.3 \times 10^4$; and (f) $L_f = 0.49$.

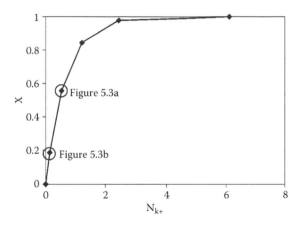

FIGURE 5.4   Effect of N_{k+} on conversion for Case I. The circled points correspond to the chromatograms in Figures 5.3a and 5.3b.

or as N_{k+} increases, the conversion X increases (Figure 5.4). At $N_{k+} \geq 5$, X is about 1.0. To increase N_{k+}, one may increase the column length, reduce the velocity, or increase k_+ by increasing the temperature.

The kinetic parameters obtained from overlapping peaks are inaccurate. Hence, good peak resolution is needed. The following factors can improve the resolution: (1) decreasing wave spreading by reducing mass transfer resistances; (2) increasing selectivity; or (3) decreasing the loading amount.

Figure 5.3d shows a simulated chromatogram in which the particle size is 10 times as large as that in Figure 5.3a. The reactant and product peaks are less well resolved than those in Figure 5.3a. As the particle size increases, the value of $1/Pe_b$ increases, and N_D and N_f decrease. Since $15N_D \sim Pe_b < N_f$, both intraparticle diffusion and axial dispersion control the peak spreading. To improve the resolution, one should increase the value of N_D and Pe_b by reducing the particle size or velocity or by increasing the column length.

Figure 5.3e shows a chromatogram as the selectivity is doubled from that in Figure 5.3a by decreasing the Langmuir "a" value of the product by 50%. When the selectivity increases, the product peak is better separated from the reactant peak, leading to a smaller overlap. In this case, the resolution is improved, but the product concentration is lower because the product migrates farther ahead of the reactant peak, increasing the width of the product peak.

Figure 5.3f shows a significant decrease in the peak resolution when the loading amount is increased 100-fold from that in Figure 5.3a. The loading factor L_f increases from 0.0049 to 0.49. At such a high loading, the column length becomes insufficient for separating the reactant and the product peaks. In general, in analytical high-performance liquid chromatography (HPLC), the loading factor should be kept less than 0.01 to ensure that the column is sufficiently long to resolve the product and the reactant peaks.

5.3.3 COLUMN DYNAMICS

To show the transient behavior of peak migration in the column in the presence of the reaction, the dynamic concentration profiles in the column of Figure 5.3a are shown in Figure 5.5. After the reactant pulse enters the column (Figure 5.5a), the reaction starts generating the product, resulting in a growing product peak, which migrates ahead of the reactant peak since it has a lower adsorption affinity (Figures 5.5b–5.5e). In the meantime, the reactant and product peaks become increasingly spread out during the migration

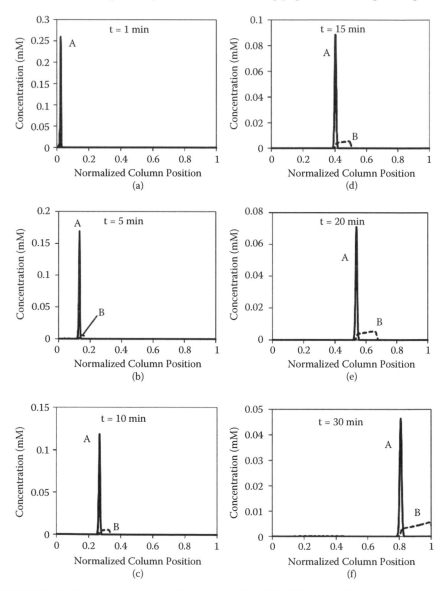

FIGURE 5.5 Dynamic column profiles generated by VERSE for Case I. The parameters are the same as in Figure 5.3a.

along the column. Eventually, a wide product peak is eluted earlier than the reactant peak (Figure 5.5f). Since the reaction is irreversible, the product keeps being formed as long as there is reactant in the column. As a result, there is always an overlapping region between the reactant and product peaks, causing some errors in determining the peak areas. As discussed in Section 5.3.2, the accuracy can be improved by increasing the resolution.

5.3.4 SUMMARY

The esterification of tetrachloroterephthaloyl chloride in a chromatography column was modeled and simulated. The results agree with the experimental data reported by Bolme and Langer [1]. The conversion of the irreversible first-order reaction increases with increasing value of the dimensionless group N_{k+}, which is the ratio of the reaction rate to the convection rate. When N_{k+} is greater than 5, the conversion approaches 1.0. Since the reactant continues to generate the product in an irreversible reaction, the reactant and product peaks always overlap. The resolution of these peaks is affected by mass transfer rates, selectivity, and the loading amount. When the particle size increases, the dimensionless axial dispersion rate ($1/Pe_b$) increases, and the film mass transfer rate N_f and intraparticle diffusion rate N_D decreases, resulting in a lower resolution. When the selectivity increases, the product peak migrates farther apart from the reactant peak, resulting in a better peak resolution but a lower product concentration. For a fixed column volume, a larger pulse gives a larger loading factor L_f. As L_f increases to 0.1 or higher, the column length is too small to resolve the reactant and product peaks, resulting in poor resolution. In general, to obtain good peak resolution in HPLC, L_f should be less than 0.01, and the values of $15N_D$, Pe_b, and N_f should be larger than 10^4.

5.4 CASE II: FIRST-ORDER REVERSIBLE REACTION

5.4.1 SYSTEM CONSIDERED AND MODEL TESTING

In the denaturation of proline-containing proteins, *cis-trans* peptide isomerization was found to be the slowest step. Since the *cis* isomer has a larger hydrophobic surface area, it can be separated from the *trans* isomer in reversed-phase chromatography. Horvath and coworkers studied the effect of slow isomerization kinetics on the peak splitting of L-alanyl-L-proline dipeptides. They found that the peaks of the *cis* and the *trans* isomers tend to merge at low flow rates or high temperatures [14,53,54]. The conditions leading to peak merging and the methods to resolve the two isomer peaks are discussed in this section.

The reaction and adsorption considered in the model are shown in Figure 5.1, where A is the *cis* isomer, and B is the *trans* isomer; k_+ and k_- are the forward and the backward reaction rate constants, respectively. The reaction is at equilibrium before the mixture enters the column, and the equilibrium constant $K = k_+/k_-$ is related to the equilibrium fractional conversion X_{eqm} by the following equation:

$$\frac{X_{eqm}}{1 - X_{eqm}} = K \tag{5.18}$$

If the peaks of A and B are well resolved, one can estimate K from the peak ratio.

Jacobson et al. [14] reported the isomerization rate constants for the *cis* and the *trans* dipeptides in the mobile phase and the stationary phase. Here, we examine a simplified case for which the reaction is assumed to occur only in the mobile phase. The effective rate constants in the mobile phase were estimated by matching simulated chromatograms with the data. The parameters used for the simulations are listed in Table 5.5. The results in Figure 5.6 agree closely with the data reported by Jacobson et al. [14].

TABLE 5.5
Simulation Parameters and Dimensionless Groups for Case II

System Parameters

L (cm)	ID (cm)	R (µm)	ε_b	ε_p
25	0.46	2.5	0.35	0.28

Operating Conditions

Figure 5.3	F (mL/min)	$C_{f,A}$ (mM)	t_l (min)
(a) and (c)	1.5	1	0.013
(b)	0.4	1	0.05

Reaction Parameters

T(°C)	k+ (min^{-1})	k_ (min^{-1})
22.5	0.766	0.418
40	4.872	2.68

Isotherm Parameters (Langmuir)

Component	a	b (mM^{-1})
A	0.9	0
B	0.33	0

Mass Transfer Parameters

Component	D_b (cm^2/min)	D_p (cm^2/min)	k_f (cm/min)	E_b (cm^2/min)
A and B	0.001	0.0005	From [37]	From [36]

Numerical Parameters

		Collocation Points		Tolerance	
Axial Elements	Step Size (L/u_0)	Axial	Particle	Absolute (m/M)	Relative
100	0.01	4	2	10^{-4}	10^{-4}

Dimensionless Groups

Figure 5.6	N_{k+}	N_{k-}	N_D	Pe_b	N_f	L_f
(a)	0.74	0.41	4×10^3	2.9×10^4	2.2×10^5	0.0052
(b)	2.78	1.52	1.5×10^4	2.9×10^4	5.4×10^5	0.0052
(c)	4.72	2.6	4×10^3	2.9×10^4	2.2×10^5	0.0052

5.4.2   Improvement of Resolution

As shown in Figure 5.6, when the temperature was high or the flow rate was low, the reactant and the product peaks merged into a single broad peak with an intermediate retention time (Figure 5.6c). The reason is that, under these conditions, the relative reaction rates to the convection rate, or N_{k+} and N_{k-}, were high. To resolve the two isomers

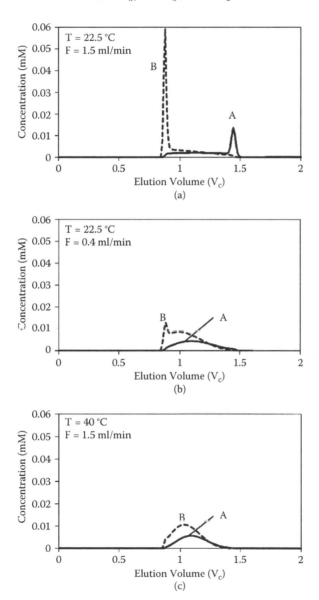

FIGURE 5.6   Simulated chromatograms for model testing for Case II. A is the *cis* isomer and B is the *trans* isomer. The temperature and the flow rate are given in the figures. Other parameters and dimensionless groups are listed in Table 5.5.

for quantitative analysis, one should decrease N_{k+} and N_{k-} and reduce the dimensionless mass transfer resistances.

Figures 5.7a–5.7c show how the peak shapes and retention times varied with the reaction rates at a fixed convection rate. The values of k_+ and k_- increased from 5.7a to 5.7c, but the equilibrium constant K remained the same. When N_{k+} and N_{k-} were about 0.1 (Figure 5.7a), the reaction rates were much slower than the convection rate. The two isomers were eluted as independent species, and the peaks were well resolved. When N_{k+} and N_{k-} were about 1 (Figure 5.7b), the reaction rates were comparable to the convection rate. The peaks of A and B overlap because the two species could convert to each other relatively fast during the migration. When N_{k+} and N_{k-} were about 10 (Figure 5.7c), the reaction rates were much larger than the convection rate. The peaks of A and B merge into a broad peak with a retention time between the two intrinsic retention times of A and B.

To increase the peak resolution, instead of decreasing the reaction rate constants, one may also increase the flow rate, thus decreasing the values of N_{k+} and N_{k-} to 0.1 or less (Figures 5.7d–5.7f). However, this strategy only works for systems with small particles. If the particle radius R is 25 μm, which is 10 times as large as the actual size in the HPLC experiments, the mass transfer resistances are large. In this case, increasing the flow rate will cause significant peak spreading, which decreases the resolution (Figure 5.8). When N_{k+} and N_{k-} are reduced to about 0.1 by increasing the flow rate, the two peaks show significant overlap since N_D is only about 10 (Table 5.6). In this case, the peak resolution is limited by intraparticle diffusion. To reduce the diffusion resistances, one should use smaller particles. For example, when $R = 2.5$ μm, N_D is greater than 10^3, and Pe_b and N_f are greater than 10^4 (Table 5.6). At these conditions, the two peaks can be resolved at high flow rates (Figure 5.7).

5.4.3 ESTIMATION OF THE EQUILIBRIUM CONSTANT

At a low reaction rate relative to the convection rate, the peaks of A and B can be well resolved. One can estimate the equilibrium constant K from the ratio of the peak areas A_B and A_A, which are proportional to the equilibrium concentrations and to the extinction coefficients ε_B and ε_A.

$$K = \frac{A_B / \varepsilon_B}{A_A / \varepsilon_A} \tag{5.19}$$

If the reaction is slow and has not reached equilibrium before the mixture enters the column, the conversion in the column at a small N_{k+} may be smaller than the equilibrium conversion. The effluent history will be similar to the ones in the irreversible reaction case (Figure 5.3), and the reactant peak will partially overlap with the product peak. To overcome this problem in estimating the equilibrium constant, one can either increase the incubation time before the sample injection or do a step change of the flow rate. For the latter case, the mobile phase velocity should be kept small at the beginning so that the reaction rates relative to the convection rate are high, and equilibrium conversion can be

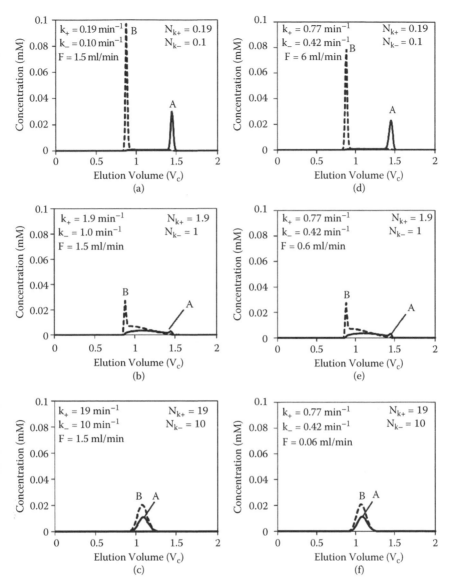

FIGURE 5.7   Effect of N_{k+} and N_{k-} on resolution for Case II with $R = 2.5$ μm. In (a), (b), and (c), N_{k+} and N_{k-} are varied by changing k_+ and k_- whereas $k_+/k_- = K = 1.9$. Other parameters are the same as in Figure 5.6a. In (d), (e), and (f), N_{k+} and N_{k-} are varied by changing the flow rate. The rate constants and loading factor are the same as in Figure 5.6a. The mass transfer dimensionless groups for (a)–(f) are listed in Table 5.6.

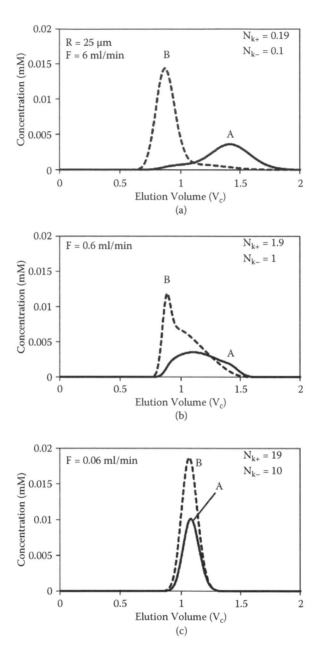

FIGURE 5.8   Effect of N_{k-} and N_{k+} on resolution for Case II with $R = 25$ μm. N_{k+} and N_{k-} are varied by changing the flow rate. The mass transfer dimensionless groups are listed in Table 5.6. Other parameters are the same as in Figure 5.6a.

TABLE 5.6

Dimensionless Mass Transfer Groups for Figure 5.7 and Figure 5.8

Figure	N_D	Pe_b	N_f
7a-7c	4×10^3	2.9×10^4	2.2×10^5
7d	1×10^3	2.9×10^4	8.8×10^4
7e	1×10^4	2.9×10^4	4.1×10^5
7f	1×10^5	2.9×10^4	1.9×10^6
8a	1×10	2.9×10^3	1.9×10^3
8b	1×10^2	2.9×10^3	8.8×10^3
8c	1×10^3	2.9×10^3	4.1×10^4

reached in the column. After the reaction equilibrium is reached, the flow rate can be increased. Then, the reaction rates are slow compared to the convection rate, and the two peaks can be separated.

5.4.4 COLUMN DYNAMICS

The dynamic column profiles for the case of peak splitting (Figure 5.7a) and peak merging (Figure 5.7c) are shown in Figures 5.9 and 5.10. When the reaction rates are slow, the peaks of *A* and *B* elute separately. When the reaction rates are high, the fast

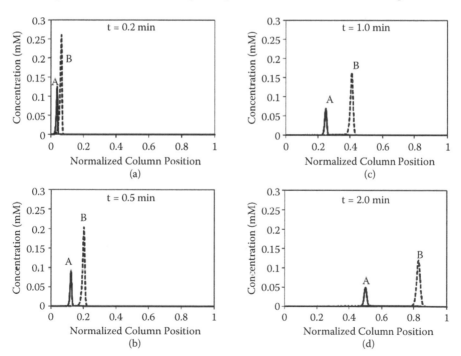

FIGURE 5.9 Dynamic column profiles for splitting peaks for Case II. The parameters are the same as in Figure 5.7a.

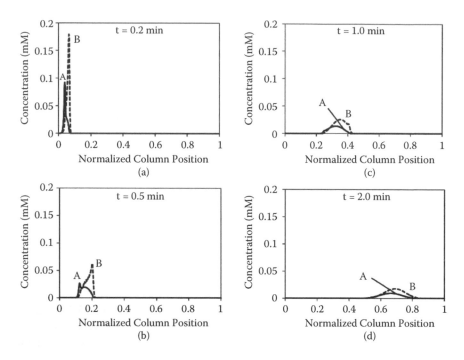

FIGURE 5.10 Dynamic column profiles for merging peaks for Case II. The parameters are the same as in Figure 5.7d.

interconversion of A and B causes a merging of the two peaks (Figure 5.10). As a result, the two isomers migrate together along the column at an intermediate speed and eventually merge into a broad peak.

5.4.5 SUMMARY

The isomerization of dipeptides was approximated as a mobile phase reaction. The simulated chromatograms were similar to the experimental chromatograms reported by Jacobson et al. [14]. The migration behavior of the two interconverting solutes depends strongly on N_{k+} and N_{k-}. When they are 0.1 or less, the reaction rates are slow compared to the convection rate, and the solutes elute as two separate peaks. When N_{k+} and N_{k-} are around 1, the two peaks overlap. When N_{k+} and N_{k-} are 10 or higher, the two peaks merge into a broad peak with a retention time between those of the pure components. To achieve a better resolution at a given temperature in systems with small particles, one needs to decrease N_{k+} and N_{k-} by increasing the velocity or decreasing the residence time. Conversely, to improve the peak resolution in a nonreactive system, one needs to increase the residence time.

For systems with large particles, the mass transfer resistances are high. Decreasing the residence time can cause significant peak spreading and poor resolution. To avoid this problem, one should use small particles to ensure that the value of N_D is greater than 10^3, and Pe_b and N_f are greater than 10^4. Under such conditions, when N_{k+} and N_{k-} are 0.1 or less and the loading factor is less than 0.01, the solute peaks can be well resolved, and then the equilibrium constant K can be estimated from the peak areas. If the reaction has not reached equilibrium before the mixture enters the column, one may force the system to reach equilibrium by increasing the incubation time before sample injection or by flow rate programming. For the latter case, the flow rate needs to be kept small at first for the reaction to reach equilibrium, and then the velocity should be high to resolve the two peaks.

5.5 CASE III: REVERSIBLE CONVERSION OF A REACTANT TO TWO PRODUCTS

5.5.1 SYSTEM CONSIDERED

Methyl formate was found to reversibly hydrolyze at room temperature to methanol and formic acid in an acidic solution. A pioneering study on this reaction in a chromatography reactor was reported by Wetherold et al. [4]. A column of activated charcoal was preequilibrated with a mobile phase of 1 M HCl, which was the catalyst for the hydrolysis reaction. The reaction and adsorption in the model are shown in Figure 5.1; A is methyl formate, B is methanol, and C is formic acid.

If a batch reactor is used for the reaction A $\leftrightarrow$ B + C, the equilibrium fractional conversion X_{eqm} is found from the equilibrium constant K and the initial reactant concentration $C_{f,A}$:

$$\frac{X_{eqm}^2}{1 - X_{eqm}} = \frac{K}{C_{f,A}} \tag{5.20}$$

The higher the reactant concentration, or the lower the value of $K/C_{f,A}$, the lower is the equilibrium conversion. If the equilibrium conversion is low, separation of the products in a chromatography column can force the reaction to reach complete conversion. At the same time, pure products are obtained.

5.5.2 MODEL TESTING

A multicomponent Freundlich isotherm model was used to correlate the adsorption isotherm data of the reactant and the products:

$$\overline{C_{p,i}} = a_i^* C_{p,i}^{m_i}, \quad i = A, B, C \tag{5.21}$$

where a^* and m are the Freundlich isotherm parameters, the values of which are listed along with those of other simulation parameters in Table 5.7. The isotherms for the components A, B, and C are plotted in Figure 5.11. Three simulated chromatograms with different loading volumes are shown in Figures 5.12a–5.12c. The predicted conversions and the shapes of the elution curves agree with the data. The parameters of Figure 5.12a were used as the benchmark in the simulations that follow.

TABLE 5.7
Simulation Parameters for Case III

System Parameters

L (cm)	ID (cm)	R (µm)	ε_b	ε_p
10.7	1	100	0.35	0.61

Operating Conditions

Figure 5.12	F (ml/min)	$C_{f,A}$ (M)	t_L (min)
(a)	0.224	1.415	107
(b)	0.222	1.43	70
(c)	0.22	1.438	18

Reaction Parameters

Reaction	k_+ (min^{-1})	k_- (M^{-1}min^{-1})
$A \leftrightarrow B + C$	0.176	0.028

Isotherm Parameters (Freundlich)

Component	a (M^{1-m})	m
A	0.38	0.428
B	0.79	0.843
C	1.46	0.357

Mass Transfer Parameters

	D_b (cm^2/min)	D_p (cm^2/min)	k_f (cm/min)	E_b (cm^2/min)
A	0.0005	0.0001	From [37]	From [36]
B	0.0005	0.0001		
C	0.0009	0.00018		

Numerical Parameters

	Step Size	Collocation Points		Tolerance	
Axial Elements	(L/u_0)	Axial	Particle	Absolute (M)	Relative
100	0.01	4	2	10^{-4}	10^{-4}

5.5.3 IMPROVEMENT OF CONVERSION

To achieve a high on-column conversion, the forward reaction rate must be sufficiently high compared to the convection rate. Otherwise, the reactant will elute rapidly before it can generate the products. To increase N_{k+}, one can either increase the residence time L/u_0 or increase the rate constant k_+ by increasing the temperature. Figure 5.13a shows the effect of N_{k+} on the conversions at three values of the ratio N_{k+}/N_{k-}, which is equal to $K/C_{f,A}$. The conversion at a given N_{k+}/N_{k-} increases with increasing N_{k+} and approaches a limiting value when N_{k+} is greater than 2. The loading factor for C, $L_{f,C} = 2.44$, is calculated based on the assumption that A is completely converted to B and C. At this high loading, the products are not well separated, and the on-column conversion, 0.87, is not much higher than the equilibrium conversion, 0.84 (Table 5.8).

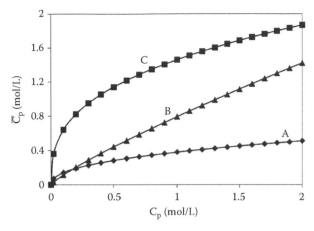

FIGURE 5.11 Freundlich isotherm curves for reactant and products for Case III. A is methyl formate, B is methanol, and C is formic acid.

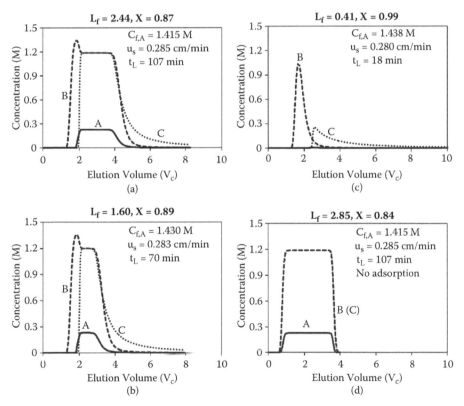

FIGURE 5.12 Simulated chromatograms for model testing for Case III. (a)–(c) Simulated chromatograms based on the operating conditions in the literature. The simulation parameters are listed in Table 5.7. (d) Hypothetical case without adsorption. The reaction and mass transfer dimensionless groups are $N_{k+} = 2.3$, $N_{k-} = 0.52$, $N_{D,A} = 26.8$, $Pe_b = 3.1 \times 10^2$, and $N_{f,A} = 1.9 \times 10^3$.

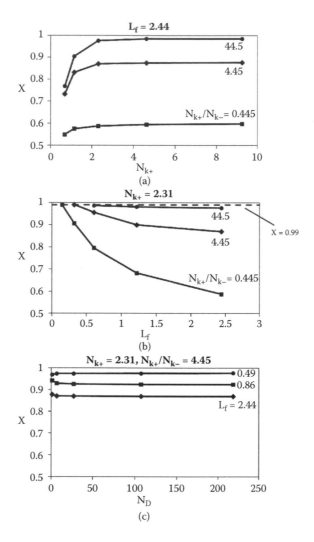

FIGURE 5.13 Effects of (a) N_{k+}, (b) L_f, and (c) N_D on conversion for Case III. In (c), N_D is varied by changing the particle size. When N_D is increased from 1.1 to 2.2×10^2, Pe_b is increased from 6.1×10 to 8.6×10^2.

TABLE 5.8

Effect of N_{k+}/N_{k-} on Equilibrium Conversion and On-Column Conversion for Case III

N_{k+}/N_{k-}	X_{eqm}	X at $L_{f,C} = 2.44$	$L_{f,C}$ for $X = 0.99$
0.445	0.48	0.60	0.15
4.45	0.84	0.87	0.3
44.5	0.98	0.99	0.6

When N_{k+} is sufficiently high (>2), one can increase the conversion by reducing $L_{f,C}$, the loading time or the loading factor for C, resulting in better separation of C from B. The on-column conversion as a function of $L_{f,C}$ for three N_{k+}/N_{k-} values is shown in Figure 5.13b. The conversions can reach 0.99 when $L_{f,C}$ is sufficiently small (Table 5.8). If N_{k+}/N_{k-} is small, the backward reaction is more favored than the forward reaction. A small loading factor is needed to separate C from B to prevent the backward reaction and achieve 0.99 conversion.

In general, a better separation of C from B results in a higher conversion. For the Freundlich isotherm model, $L_{f,C}$ is inversely proportional to a^*_C. If a^*_A is fixed and a^*_B and a^*_C are increased by the same proportion, $L_{f,C}$ becomes smaller, and the conversion is higher (Table 5.9a).

A higher sorbent selectivity for C can also achieve a better separation of C from B and thus a higher conversion (Table 5.9b). If the concentration is about 1 M, the selectivity of C to B is close to a^*_C/a^*_B for the Freundlich isotherm. If a^*_A and a^*_B are fixed and only a^*_C is increased, the selectivity will increase, and $L_{f,C}$ will decrease (Table 5.9b). The conversions are higher than those with the same $L_{f,C}$ in Table 5.9a.

If a^*_B and a^*_C are fixed and a^*_A is increased, the reactant affinity becomes higher. The reactant migrates more slowly and becomes better separated from B and C. The band of A is also more spread out and diluted because of the high value of a^*_A. These two factors result in a higher conversion according to Le Chatelier's principle. Thus, the conversion increases with increasing a^*_A (Table 5.9c).

TABLE 5.9
Effects of Product Affinity, Product Selectivity, and Reactant Affinity on Conversion for Case III

(a) Effect of Product Affinity ($a^*_A = 0.38$)

a^*_B	a^*_C	a^*_C/a^*_B	$L_{f,C}$	X
0.79	1.46	1.85	2.44	0.87
1.58	2.92	1.85	1.22	0.90
2.38	4.38	1.85	0.81	0.92
3.96	7.30	1.85	0.49	0.97

(b) Effect of Product Selectivity ($a^*_A = 0.38$)

a^*_B	a^*_C	a^*_C/a^*_B	$L_{f,C}$	X
0.79	1.46	1.85	2.44	0.87
0.79	2.92	3.7	1.22	0.91
0.79	4.38	5.6	0.81	0.96
0.79	7.30	9.3	0.49	0.99

(c) Effect of Reactant Affinity ($a^*_C/a^*_B = 1.85$)

a^*_A	a^*_B	a^*_C	$L_{f,C}$	X
0.38	0.79	1.46	2.44	0.87
1	0.79	1.46	2.44	0.88
2	0.79	1.46	2.44	0.89
5	0.79	1.46	2.44	0.93

5.5.4   TRADE-OFF BETWEEN CONVERSION AND PRODUCTIVITY

In evaluating the efficacy of a chromatography reactor, one may take into account the productivity as well as the conversion. The productivity P_r is defined as

$$P_r \equiv \frac{C_f V_f Y}{V_C \, t_{cycle}} = \frac{C_f u_{st} t_L Y}{L \, t_{cycle}} = \frac{q_f L_f Y}{t_{cycle}} \tag{5.22}$$

where Y is the yield, and t_{cycle} is the cycle time.

As discussed in Section 5.5.3, one needs to reduce L_f to achieve a high conversion. Since the productivity is proportional to L_f according to Equation (5.22), it decreases with increasing conversion (Figure 5.14). Since there is a trade-off between conversion and productivity, one has to determine the optimal loading volume, column length, and velocity based on cost or other criteria. If the reactant and the products are expensive, high conversion is more important than high productivity. Therefore, L_f needs to be small to achieve complete conversion. By contrast, if the adsorbent is costly, high productivity is important. Then, L_f should be large to achieve high productivity.

5.5.5   IMPROVEMENT OF PRODUCT SEPARATION

A great advantage of using a chromatography reactor compared to a batch reactor is that pure products can be obtained. To improve the product purity, one can (1) reduce the loading factor; (2) increase the selectivity; or (3) reduce the wave spreading.

To reduce L_f, one can either decrease the loading time (Figure 5.15b) or increase the sorbent affinities for the products (a^*_B and a^*_C, Figure 5.15c). Since both methods lower the loading factor by a factor of 5, the chromatograms and the yields of pure product B are similar. Figure 5.15d shows a further increase in the product separation

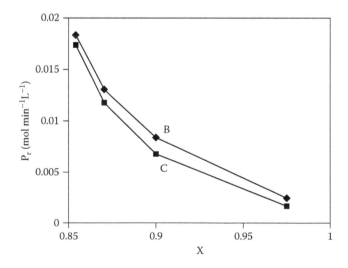

FIGURE 5.14   Trade-off between productivity and conversion for Case III.

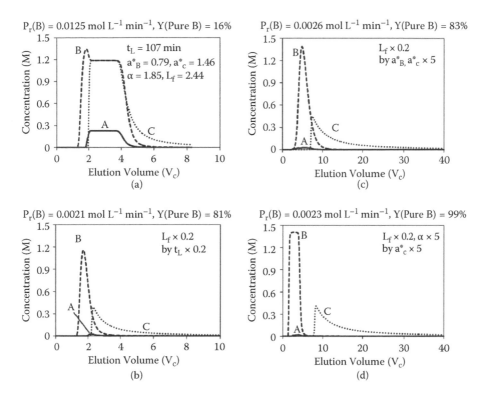

FIGURE 5.15 Effects of loading factor and selectivity on product separation for Case III. Other parameters and dimensionless groups are the same as in Figure 5.12a. The conversions are (a) 0.87, (b) 0.97, (c) 0.97, (d) 0.99.

when the selectivity is increased by a factor of 5 in addition to a fivefold decrease in L_f. Figures 5.15b–5.15d show higher conversions and product purities but lower productivities than those in Figure 5.15a.

Wave spreading due to axial dispersion and intraparticle diffusion also has an impact on product separation. Figure 5.16 shows the chromatograms obtained for different particle sizes with $L_f = 0.49$. When both N_D and Pe_b are low (Figure 5.16a), the peaks are broad, resulting in poor separation and low product concentrations. As N_D and Pe_b increase, the separation improves, and the product concentrations increase. When both N_D and Pe_b are of the order of 10^2, further gains in separation by increasing N_D and Pe_b are small (Figures 5.16c and 5.16d). The wave spreading under these conditions is caused by the nonlinearity of the isotherms rather than by the mass transfer resistances.

The conversions are nearly constant as the particle radius varies from 500 to 35 μm, but the yields of pure product B are higher for the systems with smaller particles (Figures 5.16a–5.16d). When the particle size is large (Figure 5.16a), there is a significant overlap of peaks B and C, which favors the backward reaction. However, the peaks are also broadened and diluted, resulting in a lower rate of the backward reaction. The two effects counterbalance, and the particle size has no substantial impact on the conversion. As shown in Figure 5.13c, at the high feed concentration,

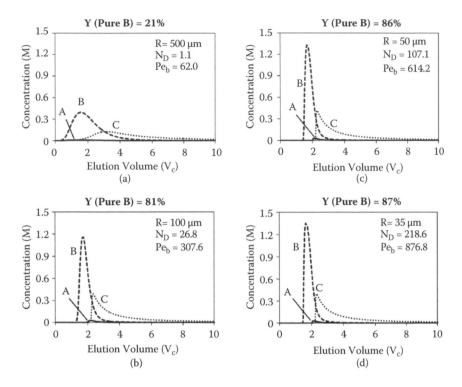

FIGURE 5.16 Effect of mass transfer resistances on separation for Case III. Different N_D and Pe_b values are obtained by varying particle sizes. $N_f = 1.3 \times 10^2$ (a); 1.9×10^3 (b); 6.0×10^3 (c); 1.1×10^4 (d). Other parameters and dimensionless groups are the same as in Figure 5.15b. The conversions are all about 0.97.

the conversion is relatively insensitive to the dimensionless mass transfer rate but is limited by the loading factor. Hence, one can use large particles to reduce the cost if high conversion is more important than high product purity or concentration.

5.5.6 COLUMN DYNAMICS

The dynamic column profiles of the system in Figure 5.12a are shown in Figure 5.17. Although A has the lowest affinity, the A band during loading does not migrate ahead of the B and C bands because of the fast conversion of A to B and C. B is displaced by C, resulting in a roll-up of B in front of the adsorption wave of C (Figure 5.17b). Once the fronts of A, B, and C are eluted before the loading ends, the reaction is at equilibrium near the column outlet (Figure 5.17c). After the loading ends, the desorption waves become diffuse, and the solutes are eluted with the affinity sequence $A < B < C$ (Figures 5.17d–5.17f).

Figure 5.18 shows the column dynamics when the adsorbent has no affinity for the solutes. The effluent history is shown in Figure 5.12d. The components B and C migrate at the same speed without separation, and finally the equilibrium conversion (0.84) is reached.

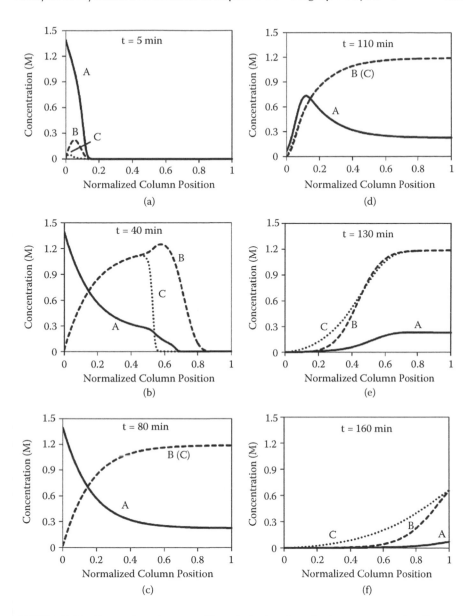

FIGURE 5.17 Dynamic column profiles of the benchmark chromatogram for Case III (Figure 5.12a).

Figure 5.19 shows the dynamics of the chromatogram in Figure 5.15d. The high selectivity, $a^*_C/a^*_B = 9.3$, and the low loading factor, $L_f = 0.49$, allow C to migrate more slowly than B and prevents the backward reaction. The concentration of A drops rapidly after the end of loading. Eventually, the conversion of A reaches 0.99, and the two products are pure. A comparison of Figures 5.17, 5.18, and 5.19 shows how the separation of the two products can improve the conversion.

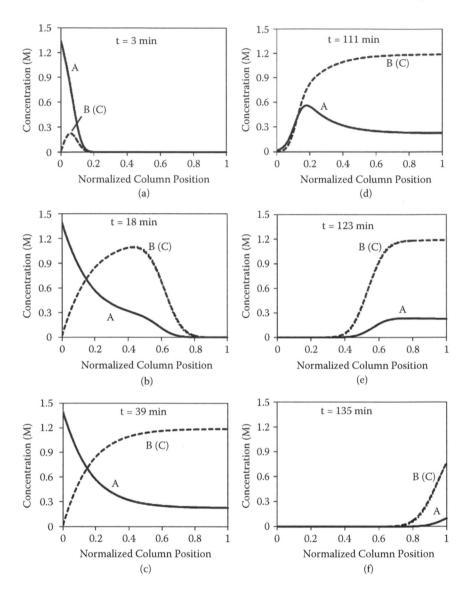

FIGURE 5.18 Dynamic column profiles of the chromatogram in Figure 5.12d, in which none of the components can be adsorb. The parameters are the same as in Figure 5.12a, except that the isotherm parameters are equal to zero.

5.5.7 SUMMARY

The hydrolysis of methyl formate catalyzed by hydrochloric acid was modeled and simulated. The simulated chromatograms agreed closely with the experimental results of Wetherold et al. [4]. A conversion higher than the equilibrium conversion at a given feed concentration can be achieved in chromatography because of the separation of the products. The on-column conversion can reach 0.99 if the relative

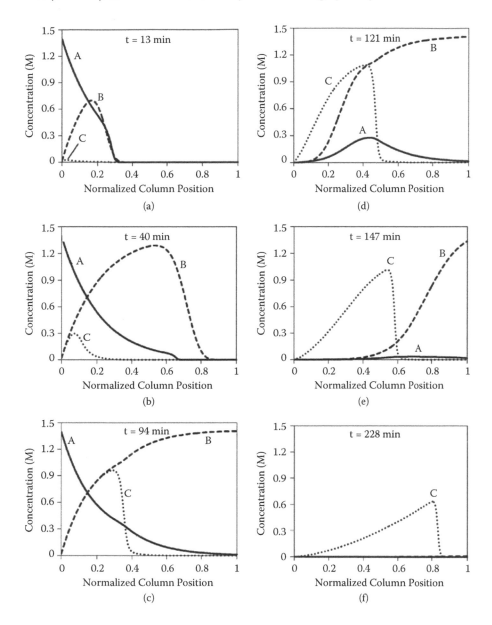

FIGURE 5.19 Dynamic column profiles of the chromatogram in Figure 5.15d, where $X = 0.99$ and the yield of pure $B = 99\%$.

reaction rate N_{k+} is more than 2 and the loading factor L_f is less than 0.3. In general, high reactant and product affinities and high product selectivities favor high conversions. A chromatography reactor can generate pure products. To improve product separation, one can reduce the loading factor L_f, which results in lower productivity. There is a trade-off between conversion and productivity. Another way to improve separation and product concentrations is to decrease the mass transfer resistances, or increase N_D and Pe_b, by using small particles or by increasing the residence times (L/u_s). When both N_D and Pe_b are larger than 10^2, tailing of the desorption waves is due to the nature of the Freundlich isotherm. Further increase of N_D and Pe_b does not improve the separation.

5.6 CASE IV(A): REVERSIBLE PROTEIN DIMERIZATION IN SIZE EXCLUSION CHROMATOGRAPHY

5.6.1 System Considered

Size exclusion chromatography (SEC) has been widely used to determine the sizes and molecular weights of proteins and polymers. A larger molecule can access a smaller fraction of the particle pore volume, or a smaller size exclusion factor Ke, and thus migrates faster than a smaller molecule. However, if proteins aggregate in solution, one may observe multiple peaks or a single broad peak, depending on the flow rate, column length, sample concentration, and temperature. Yu and coworkers studied the effect of reversible dimerization on peak shapes and elution behavior [28,55]. They found that the aggregation rate and the dissociation rate relative to the convection rate, as represented by N_{k+} and N_{k-}, are the two key dimensionless groups. A two-dimensional diagram showing the effects of these groups on the peak shape and the retention behavior was generated. It can serve as a road map for predicting the general elution behavior of proteins in SEC with reversible dimerization.

The dimerization reaction occurring in the column is described in Figure 5.1d, where M, the monomer, and D, the dimer, do not adsorb. The separation of M and D arises from the difference in the size exclusion factor Ke. The reaction is assumed to have reached equilibrium before the mixture enters the column. The fractional equilibrium conversion in a batch reactor depends on $KC_{f,M}$:

$$\frac{X_{eqm}}{(1 - X_{eqm})^2} = 2KC_{f,M} \tag{5.23}$$

where $C_{f,M}$ is the total initial monomer concentration in the feed. If the reaction rates are much slower than the convection rate, M and D can be well separated. One can measure K from the ratio of the peak areas of M and D.

5.6.2 Effects of N_{k+}, N_{k-}, and Their Ratio on the Number of Peaks, Peak Shapes, and Retention Times

Simulated chromatograms with various dimerization and dissociation rates, k_+ and k_-, respectively, are shown in a two-dimensional diagram in Figure 5.20. The parameter values in the simulations are listed in Table 5.10. The values of N_{k+} and N_{k-} range from 0.1 to 100. The monomer and dimer concentrations are normalized by $C_{f,M}$. The dimer molar concentrations are multiplied by two to reflect the monomer inventory so that one can tell the conversion from the ratio of the peak areas.

TABLE 5.10

Simulation Parameters for Generating the 2D Diagram in Case IV(A)

System Parameters

L (cm)	ID (cm)	R (μm)	ε_b	ε_p
180	5.1	54	0.35	0.89

Operating Conditions

F (mL/min)	$C_{f,M}$ (M)	t_L (min)
2.6	0.014	66.5

Reaction Parameters

Reaction	k_+ ($M^{-1}min^{-1}$)	k_- (min^{-1})
$2M \leftrightarrow D$	0.0143–14.3	0.0002–0.2

Isotherm Parameters (Size Exclusion)

Component	Ke	a	b (M^{-1})
M	0.74	0	0
D	0.25	0	0

Mass Transfer Parameters

Component	D_b (cm^2/min)	D_p (cm^2/min)	k_f (cm/min)	E_b (cm^2/min)
M	0.0000549	0.0000229	From [37]	From [36]
D	0.000048	0.00002		

Numerical Parameters

Axial Elements	Step Size (L/u_0)	Collocation Points Axial	Collocation Points Particle	Tolerance Absolute (M)	Tolerance Relative
50	0.001	4	2	10^{-4}	10^{-4}

Figure 5.20 can be generated by varying the rate constants k_+ and k_- or by varying the feed concentration C_f, the interstitial velocity u_0, or the column length L [28].

The dimensionless retention volumes for the monomer and the dimer are about 0.8 and 0.5 (Figure 5.20). For small association and dissociation rates, N_{k+} and N_{k-} are small (0.1), and the monomer and the dimer peaks are well resolved (Figure 5.20u). As N_{k+} increases, the percentage of dimer increases, and eventually only the dimer peak is observed (see the sequence u, p, k, f, a). As N_{k-} increases for $N_{k+} = 0.1$, the percentage of monomer increases (sequence u, v, w, x, y), and eventually only one monomer peak is observed.

For the cases of $N_{k+} = N_{k-}$ (the diagonal sequence u, q, m, i, e), as N_{k+} and N_{k-} increase, the two peaks start merging when N_{k+} and N_{k-} are about 1 (Figures 5.20q and 5.20m). For N_{k+} and $N_{k-} \geq 10$, monomers and dimers convert rapidly to each other during the migration. The two peaks coelute as one broad peak with a retention time between those of the monomer and the dimer (Figure 5.20i and 5.20e). A similar trend is also observed for other N_{k+}/N_{k-} values as N_{k+} and N_{k-} increase.

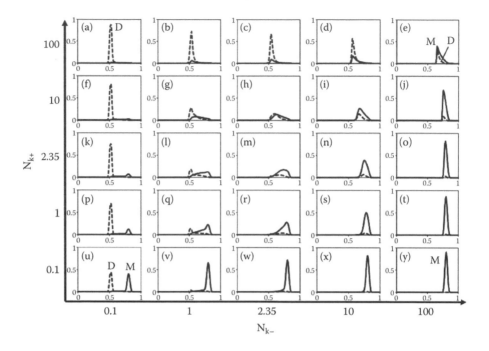

FIGURE 5.20 Elution behaviors of monomer peak $(C_M/C_{f,M})$ and dimer peak $(2C_D/C_{f,M})$ versus normalized elution volume (by column volume) at different N_{k+} and N_{k+} for Case IV (A). The solid lines are for the monomer, and the dashed lines are for the dimer. The parameters used in the simulations are listed in Table 5.10. For all cases, $N_{D,M} = 4.8 \times 10^2$, $Pe_b = 9.6 \times 10^3$, $N_{f,M} = 2.4 \times 10^4$, and $L_f = 0.047$.

The ratio N_{k+}/N_{k-}, which is equal to $KC_{f,M}$, is the dimerization rate relative to the dissociation rate. For the chromatograms in the diagonal sequence, $N_{k+}/N_{k-} = 1$, both the monomer and the dimer peaks are similar in size. When N_{k+} and N_{k-} are larger than 1, the monomer peak is slightly larger than the dimer peak (Figures 5.20q, 5.20m, 5.20i, and 5.20e). This is due to broadening and dilution of the monomer peak by the rapid reactions. As the monomer concentration C_M decreases, the dimer concentration C_D decreases more rapidly since $C_D = KC_M^2$, resulting in a larger monomer peak.

For $N_{k+}/N_k < 1$, the chromatograms below the diagonal sequence, the dimer peak is smaller than the monomer peak. For $N_{k+}/N_k > 1$, the chromatograms above the diagonal sequence, the dimer peak is larger than the monomer peak, except Figures 5.20l and 5.20h. The monomer peak is slightly larger than the dimer peak because the peak broadening due to the reactions favors the formation of monomers.

5.6.3 Estimation of the Equilibrium Constant

For small N_{k+} and N_{k-} (about 0.1), the reaction rates are slower than the convection rate. The distribution of monomers and dimers remains equal to the initial equilibrium distribution during the migration, and the peaks are well separated. Then, the equilibrium constant can be estimated from the peak areas. If one assumes that the extinction coefficient of the dimer is twice that of the monomer, then the peak area A_M of the monomer is proportional to its concentration C_M, and the peak area of the dimer A_D is proportional to $2C_D$.

$$C_M = C_{f,M}\left(\frac{A_M}{A_D + A_M}\right)$$ (5.24)

Since $C_D = KC_M^2$ and $C_M + 2C_D = C_{f,M}$, one obtains

$$K = \frac{C_{f,M} - C_M}{2C_M^2}$$ (5.25)

5.6.4 SUMMARY

Simulations were used to show the effects of N_{k+}, N_{k-}, and their ratio on peak shapes and retention times in SEC. A two-dimensional diagram with the chromatograms at different N_{k+} and N_{k-} values can be used to illustrate the elution behavior of the monomer and dimer peaks. If N_{k+} and N_{k-} are small (~0.1 or less), the dimerization and dissociation rates are slow compared to the convection rate. Then, the monomers and dimers behave as independent components, and they are eluted separately. Under such conditions, the distribution of monomers and dimers remains the same as the equilibrium distribution, and the equilibrium constant K can be estimated from the relative peak areas. If N_{k+} and N_{k-} are large (>1), then the dimerization and dissociation rates are high. The monomer and dimer peaks tend to overlap and eventually merge into a single broad peak with a retention time between those of the pure monomer and the pure dimer peaks. When N_{k+}/N_{k-} is 10 or greater, the dimer peak is dominant. When N_{k+}/N_{k-} is 0.1 or less, the monomer peak is dominant. When N_{k+}/N_{k-} is about 1, both the monomer and dimer peaks are seen. The dimer peak is smaller than the monomer peak when N_{k+} and N_{k-} are greater than 1 because the fast reversible reaction causes merging and broadening of the two peaks. The reduction in peak concentration favors the dissociation of the dimer to the monomer.

5.7 CASE IV(B): REVERSIBLE DIMERIZATION OF A PROTEIN IN IMMOBILIZED METAL AFFINITY CHROMATOGRAPHY

5.7.1 SYSTEM CONSIDERED AND MODEL TESTING

Immobilized metal affinity chromatography (IMAC) is widely used for selective protein capture. Proteins containing histidine, for example, can form coordination bonds with specific metal ions, such as copper, nickel, and cobalt. These proteins can be isolated from others using an IMAC column. The adsorbed protein can be eluted by adjusting pH, changing solvent strength, or adding a competitive displacer.

TABLE 5.11
Simulation Parameters for Case IV (B)

System Parameters

L (cm)	ID (cm)	R (µm)	ε_b	ε_p
7.5	1	5	0.38	0.81

Operating Conditions

Figure 5.21	F (mL/min)	$C_{f,M}$ (M)	t_L (min)
(a)	1.778	5.94×10^{-5}	200
(b)	1.778	8.87×10^{-5}	200
(c)	1.778	1.477×10^{-4}	200

Reaction Parameters

Reaction	k_+ (M⁻¹min⁻¹)	k_- (min⁻¹)
$2M \leftrightarrow D$	240	0.2238

Isotherm Parameters (Langmuir)

Component	a	b (M⁻¹)
M	85	15,000
D	10	300,000

Mass Transfer Parameters

Component	D_b (cm²/min)	D_p (cm²/min)	k_f (cm/min)	E_b (cm²/min)
M	0.000066	3.7×10^{-7}	From [37]	From [36]
D	0.000033	3.7×10^{-8}		

Numerical Parameters

Axial Elements	Step Size (L/u_0)	Collocation Points		Tolerance	
		Axial	Particle	Absolute (M)	Relative
100	0.01	4	2	10^{-7}	10^{-4}

Whitley et al. [16] studied the aggregation effect on the breakthrough curves of myoglobin in IMAC. They assumed that only reversible dimerization of myoglobin takes place in the mobile phase, and that the adsorption of the monomer and the dimer follows the nonlinear Langmuir adsorption isotherm. Their simulated frontal chromatograms based on the hypothesis of mobile phase reversible dimerization agree well with the experimental data by Wong [56]. The parameters used in the simulations are given in Table 5.11. Here, we plot modified simulated chromatograms in Figure 5.21, in which the dimer concentrations are twice the molar concentration.

The dimerization occurring in the column assumed in the simulations is described in Figure 5.1, where M is the monomer and D is the dimer. The reaction is assumed to have reached equilibrium before the mixture enters the column. The equilibrium conversion is determined by the value of $KC_{f,M}$.

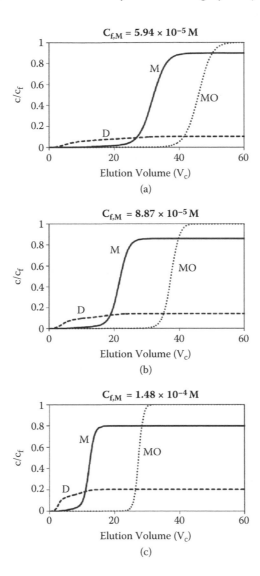

FIGURE 5.21   Simulated chromatograms at different feed concentrations for Case IV (B). *M* is the monomer, and *D* is the dimer. The curves marked as *MO* (monomer only) are for the pure monomer. The simulation parameters are listed in Table 5.11. $N_{k+} = 0.018$ (a); 0.027 (b); 0.045 (c). Other dimensionless groups are $N_{k-} = 0.282$; $N_{D,M} = 2.46$; $Pe_h = 4.0 \times 10^3$; $N_{f,M} = 76 \times 10^3$.

5.7.2 Effect of Adsorption Isotherm, N_{k+}, N_{k-}, and N_{k+}/N_{k-} on the Breakthrough Curves

In Figure 5.21, the dimerization and the dissociation rates are slow compared to the convection rate; $N_{k+} < 1$ and $N_{k-} < 1$. The monomer and the dimer migrate as independent species, resulting in two breakthrough curves. The dimer front is ahead of the monomer front because the dimer has a larger Langmuir "b" value and a smaller

Langmuir "a" value, resulting in a small capacity or an early breakthrough in the presence of the monomer.

The curves marked *MO* in Figure 5.21 are the breakthrough curves for the pure monomer if no aggregation occurs. The breakthrough times of the monomer in the reactive system are significantly shorter than for a nonreactive system because the dimer with a large value of "bC" can occupy a significant fraction of the binding sites, resulting in a greatly reduced monomer capacity.

As the feed concentration $C_{f,M}$ increases from 5.94×10^{-5} M to 1.48×10^{-4} M (Figures 5.21a–5.21c), the ratio N_{k+}/N_{k-}, which is equal to $KC_{f,M}$, increases from 0.06 to 0.16, and the conversion of the monomer to the dimer increases from 0.1 to 0.2. The breakthrough times of the monomer and the dimer are shorter because of the higher feed concentration.

If K and $C_{f,M}$ are the same as in Figure 5.21a, but k_+ and k_- vary while their ratio is constant, N_{k+} and N_{k-} change accordingly, but the ratio N_{k+}/N_{k-} is constant and equal to 0.06. The simulated frontal curves at various dimensionless reaction rates are shown in Figure 5.22. When N_{k+} and N_{k-} are less than 1, the reaction rates are slow compared to the convection rate (Figures 5.22a and 5.22b). Then, the dimers and the monomers migrate as independent species. When N_{k-} is greater than 1, the

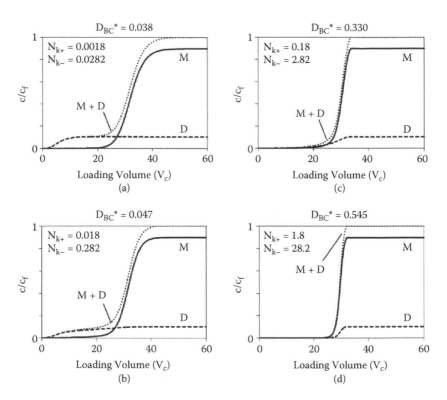

FIGURE 5.22 Frontal curves at various reaction rates for Case IV (B). N_{k+} and N_{k-} are varied by changing k_+ and k_-; $k_+/k_- = K = 1072.4$ M^{-1}. Other parameters and dimensionless groups are the same as in Figure 5.21a.

dissociation rate increases, and the dimer front starts to merge with the monomer front (Figure 5.22c). When N_{k-} is greater than 10, the dimer front merges totally with the monomer front (Figure 5.22d).

5.7.3 Dynamic Binding Capacity

When N_{k+} and N_{k-} are smaller than 1, the dimer front migrates faster than the monomer front, resulting in an early breakthrough, which is undesirable. When the breakthrough concentration is only 1% of the feed concentration, little protein is lost in the capture step. For this reason, the dynamic binding capacity D_{BC} is defined as the amount of protein captured per milliliter column volume when the breakthrough concentration C_{br} reaches 1% of the feed concentration $C_{f,M}$. Furthermore, D_{BC} is normalized by q_f, the solid phase concentration in equilibrium with the feed concentration. This ratio is the dimensionless dynamic binding capacity $D_{BC}{}^*$, which is equal to the loading factor (see Section 5.2.4), at which the breakthrough concentration is 1% of the feed concentration:

$$D_{BC}^* \equiv \frac{D_{BC}}{q_f} = L_f(C_{br} = 0.01C_{f,M}) = \frac{C_{f,M}V_f(C_{br} = 0.01C_{f,M})}{q_f V_C} \qquad (5.26)$$

A high value of $D_{BC}{}^*$, for example, $D_{BC}{}^* = 0.95$, indicates that a large fraction, 95%, of the sorbent is used for the protein capture.

The values of $D_{BC}{}^*$ increase with increasing N_{k+} and N_{k-} (Figures 5.22 and 5.23) since the dimer front merges with the monomer front when the reaction rates are high. $D_{BC}{}^*$ increases with increasing N_{k-} and approaches a plateau for N_{k-} greater than 10. When the feed concentration is high (5.94×10^{-4} M), the conversion of monomer to dimer is high (0.41). The value of the quantity $(1 + b_M C_M + b_D C_D)/(1 + b_M C_{f,M})$ is large (4.43), indicating that the presence of dimer reduces significantly the amount of monomer that can adsorb. There is a big difference between the limiting values of $D_{BC}{}^*$ with and without dimerization. There is a similar trend for $C_{f,M} = 5.94 \times 10^{-5}$ M, for which the conversion is 0.1. For $C_{f,M} = 5.94 \times 10^{-6}$ M, the conversion is only 0.01 and $(1 + b_M C_M + b_D C_D)/(1 + b_M C_{f,M}) = 1.01$, the dimer has little effect on the adsorption of the monomer. In this case, the limiting $D_{BC}{}^*$ approaches the limiting value without dimerization. This value is limited mainly by wave spreading due to intraparticle diffusion. A higher feed concentration gives a higher $D_{BC}{}^*$ for the monomer without reaction because the wave front is self-sharpened when the feed concentration is in the nonlinear isotherm region, where $b_M C_{f,M} \sim 1$ or $\gg 1$.

In addition to increasing the relative reaction rates, decreasing mass transfer resistances can increase $D_{BC}{}^*$ by reducing wave spreading. Since the protein has a very low diffusivity, N_D is three orders of magnitude smaller than Pe_b and N_f, indicating that the intraparticle diffusion controls wave spreading. Figure 5.24 shows the chromatograms obtained at different N_D values for various particle sizes. The values of K and $C_{f,M}$ are the same as those in Figure 5.21a. N_{k+} and N_{k-} are the same as those in Figure 5.22d. When N_D increases, the monomer and the dimer fronts become sharper, resulting in a higher $D_{BC}{}^*$. Figure 5.25 shows that $D_{BC}{}^*$ increases with increasing

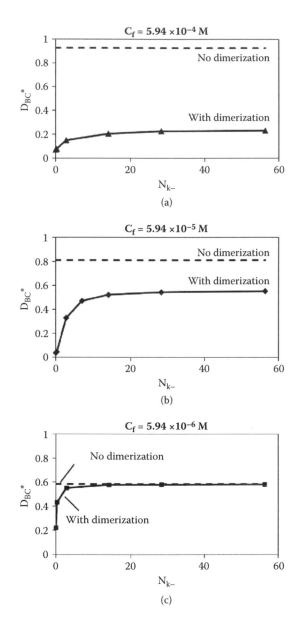

FIGURE 5.23 Effect of N_{k-} on dimensionless dynamic binding capacity D_{BC}^* at various feed concentrations for Case IV (B). $K = k_+/k_- = 1072.4$ M^{-1}. The horizontal dashed lines are the D_{BC}^* of the monomer without dimerization. N_D, Pe_b and N_f are the same as in Figure 5.21a.

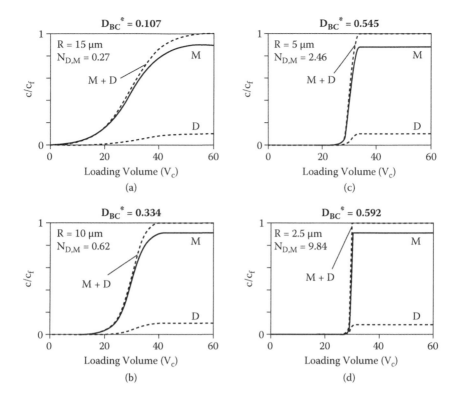

FIGURE 5.24 Frontal curves at different mass transfer resistances for Case IV(B). The particle size and $N_{D,M}$ are shown in each figure. $Pe_b = 1.3 \times 10^3$ (a); 2.0×10^3 (b); 4.0×10^3 (c); 7.9×10^3 (d). $N_{f,M} = 1.2 \times 10^3$ (a); 2.4×10^3 (b); 7.6×10^3 (c); 2.4×10^4 (d). Other parameters are the same as in Figure 5.22d.

N_D at various feed concentrations. When $C_{f,M}$ is relatively high (Figures 5.25a and 5.25b),$_{,BC}$ D_{BC}^* rapidly reaches the limiting value. The gaps between the limiting values of D_{BC}^* and that of the monomer without dimerization are due to the decreasing adsorption capacity of the monomer in the presence of the dimer. When $C_{f,M}$ is low (Figure 5.25c), the adsorption of the dimer has little effect on the adsorption of the monomer, and the two curves in Figure 5.25c merge. The approach to the limiting value is slower than those in Figures 5.25a and 5.25b because at such a low feed concentration, there is no self-sharpening effect.

Although a low feed concentration ($\sim 10^{-6} M$) results in a high value of D_{BC}^*, the actual dynamic binding capacity D_{BC} is low because the equilibrium binding capacity q_f is small [see Equation (5.26)]. The D_{BC} value in Figure 5.25b at N_D about 10 is higher than those in Figures 5.25a and 5.25c. Thus, to maximize the capture amount for a given column, one should keep N_{k-} greater than 10 to merge the fronts, N_D greater than 10 to sharpen the fronts, and an appropriate feed concentration to maximize the dynamic binding capacity.

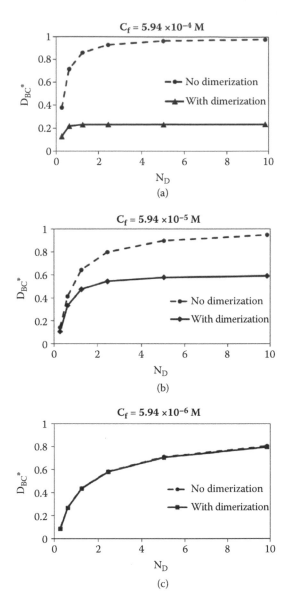

FIGURE 5.25 Effect of N_D on $D_{BC}{}^*$ at different feed concentrations for Case IV(B). $K =$ 1072.4 M^{-1}. The dashed lines are the $D_{BC}{}^*$ of the monomer without dimerization. $N_{k-} = 28.2$ as in Figure 5.22d. $N_{k+} = 18$ (a); 1.8 (b); and 0.18 (c).

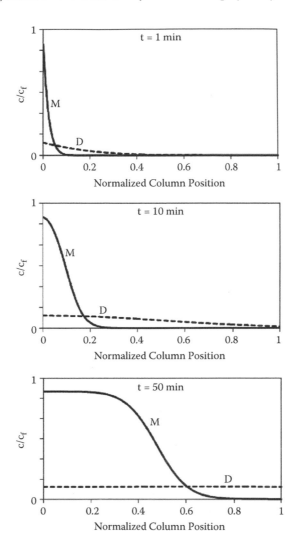

FIGURE 5.26 Dynamic column profiles of the chromatogram in Figure 5.22a. Two plateaus are formed at relatively low reaction rates.

5.7.4 COLUMN DYNAMICS

The dynamic column profiles for Figure 5.22a with two fronts and Figure 5.22d with a merged front are plotted in Figures 5.26 and 5.27, respectively. When the dimerization and dissociation rates are slow, as in Figure 5.26, the dimer front moves faster than the monomer front, resulting in two plateaus. When the reactions rates are high, once the dimer front moves ahead of the monomer front, the dimer will dissociate rapidly. Then, the dimer front merges spontaneously with the monomer front, and the two species elute together, as in Figure 5.27. In both cases, the migration speed of the monomer is larger than that of the monomer alone because the dimer occupies a

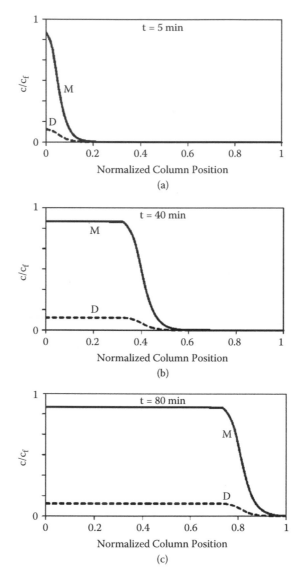

FIGURE 5.27 Dynamic column profiles of the chromatogram in Figure 5.22d. Only one plateau with merged monomer and dimer fronts is formed at high reaction rates.

significant fraction of the adsorption sites and reduces the fraction of the sites available for monomer adsorption.

5.7.5 Summary

The capture of myoglobin in the presence of reversible aggregation using an IMAC column has been simulated and analyzed. The simulated frontal curves based on the assumption of mobile phase dimerization agreed with the experimental data [56].

Since the dimer has a much lower value of the Langmuir "a" parameter, it migrates faster than the monomer in the column. The frontal curves can have two plateaus when N_{k+} and N_{k-} are small (~0.1 or less). The early breakthrough of the dimer reduces greatly the dimensionless dynamic binding capacity D_{BC}^*. To merge the dimer front with the monomer front and increase D_{BC}^*, one can increase N_{k+} and N_{k-}. One can also increase the relative intraparticle diffusion rate N_D for reducing wave spreading. If the feed concentration is relatively high (10^{-5}–10^{-4} M), the maximum value of D_{BC}^* is lower than that of the monomer without dimerization. This difference is due to the presence of the dimer, which occupies a significant fraction of the adsorption sites, and reduces the adsorption capacity for the monomer. When the feed concentration is low (~10^{-6} M), the dimer concentration is too low to have any effect on the adsorption of the monomer. Thus, D_{BC}^* is similar to that of the monomer alone and can approach 1 if N_{k+}, N_{k-}, and N_D are sufficiently high. To maximize the capture amount for a given column, one should keep both N_{k-} and N_D greater than 10.

5.8   CASE V: MULTIPLE AGGREGATION REACTIONS IN HYDROPHOBIC INTERACTION CHROMATOGRAPHY

5.8.1   System Considered and Model Testing

The previous two sections showed the effects of dimerization on SEC and IMAC. In certain cases, however, a protein or a polymer can form higher-order aggregates, resulting in more than two peaks in elution chromatography or more than two fronts in frontal chromatography. Grinberg et al. [3] studied the effect of the temperature on the behavior of β-lactoglobulin A in hydrophobic interaction chromatography (HIC) and found that the primary aggregates are tetramers, octamers, and dodecamers. Van Cott et al. [15] used a set of reaction parameters to simulate the peak-splitting and peak-merging phenomena at various temperatures and flow rates. Here, it is assumed that the reaction equilibrium constants are fixed, and the temperature affects only the reaction rates.

The two reactions occurring in the column are described in Figure 5.1, where T is the tetramer, O is the octamer, and D is the dodecamer. The reactions are assumed to have reached equilibrium before the mixture enters the column, and the distribution of the aggregates depends on the reaction equilibrium constants K_1 and K_2 and the total concentration in the feed $C_{f,T}$. If the reaction rates are slow and the peaks are well resolved, one can estimate the reaction stoichiometry and the reaction equilibrium constants from the peak ratios at several feed concentrations.

In simulating the elution chromatograms, isocratic elution is assumed. In the experiments of Grinberg et al. [3], decreasing salt gradient elution was used. The adsorption isotherm parameters were adjusted to match the simulated retention times with the experimental data. The adsorption of each aggregate (tetramer, octamer, and dodecamer) was assumed to follow the nonlinear Langmuir adsorption isotherm. The parameters used in the simulations are given in Table 5.12. Simulated chromatograms for different temperatures are shown in Figures 5.28a–5.28d, which are similar to Figure 5.5 of Grinberg et al. [3]. The octamer and dodecamer concentrations shown in the figures were two and three times larger than their actual molar concentrations so that the peak ratios reflected the conversion of tetramers to

TABLE 5.12
Simulation Parameters and Dimensionless Groups for Case V

System Parameters

L (cm)	ID (cm)	R (μm)	ε_b	ε_p
10	0.46	2.5	0.38	0.65

Operating Conditions

F (ml/min)	$C_{f,M}$ (M)	t_L (min)
1	2.717×10^{-4}	4.23×10^{-3}

Reaction Parameters

T (°C)	$k_{+,1}$ (M^{-1}min^{-1})	$k_{-,1}$ (min^{-1})	$k_{+,2}$ (M^{-1}min^{-1})	$k_{-,2}$ (min^{-1})
4	7,998	0.3334	1,500	0.4546
10	15,000	0.6252	1,850	0.5606
15	22,002	0.9168	7,002	2.121
25	96,000	4	13,200	4

Isotherm Parameters (Langmuir)

	Tetramer		Octamer		Dodecamer	
T (°C)	a	b (M^{-1})	a	b (M^{-1})	a	b (M^{-1})
4	6.70	500	7.48	500	6.00	500
10	6.95	500	7.80	500	6.30	500
15	7.14	500	7.82	500	6.38	500
25	7.40	500	7.92	500	6.40	500

Mass Transfer Parameters

Component	D_b (cm^2/min)	D_p (cm^2/min)	k_f (cm/min)	E_b (cm^2/min)
T	0.002154	0.000769	From [37]	From [36]
O	0.0004314	0.000154		
D	0.0004314	0.000154		

Numerical Parameters

		Collocation Points		Tolerance	
Axial Elements	Step Size (L/u_0)	Axial	Particle	Absolute (M)	Relative
100	0.01	4	2	10^{-7}	10^{-4}

Dimensionless Groups

Figure 5.28	$N_{k+,1}$	$N_{k-,1}$	$N_{k+,2}$	$N_{k-,2}$	$N_{D,T}$	Pe_b	$N_{f,T}$	L_f
(a)	1.373	0.211	0.257	0.287	8.2×10^3	1.1×10^4	1.7×10^5	0.0020
(b)	2.574	0.395	0.317	0.354	8.2×10^3	1.1×10^4	1.7×10^5	0.0020
(c)	3.775	0.579	1.201	1.339	8.2×10^3	1.1×10^4	1.7×10^5	0.0019
(d)	16.47	2.525	2.266	2.525	8.2×10^3	1.1×10^4	1.7×10^5	0.0018

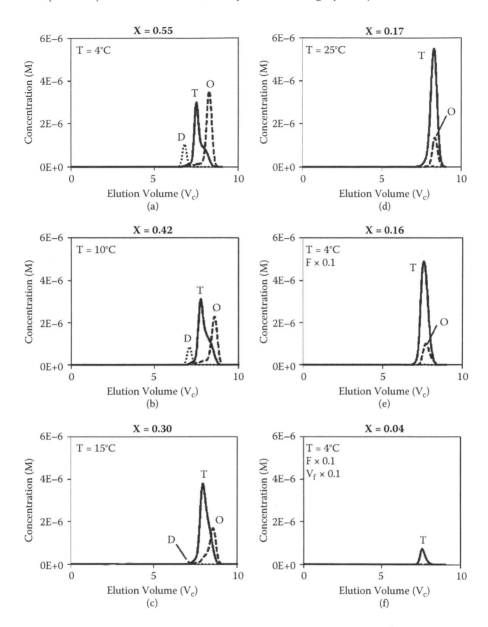

FIGURE 5.28 Simulated chromatograms for Case V for various temperatures, flow rates, and injection volumes. T is the tetramer, O is the octamer, and D is the dodecamer. Cases (a)–(d) are used for testing the models. The parameters and the values of dimensionless groups are listed in Table 5.12. Cases (e) and (f) have the same temperature but different flow rate and loading amount compared to (a). The dimensionless groups are $N_{k+,1} = 13.73$; $N_{k-,1} = 2.11$; $N_{k+,2} = 2.57$; $N_{k-,2} = 2.87$; $N_{D,T} = 8.2 \times 10^4$; $Pe_b = 1.1 \times 10^4$; $N_{f,T} = 7.9 \times 10^5$; and $L_f = 0.002$ (e); 0.0002 (f).

octamers and dodecamers. The thus-determined reaction and adsorption parameters were used in the following simulations.

5.8.2 PEAK SPLITTING, PEAK MERGING, AND COLUMN DYNAMICS

At 4°C, the tetramer and octamer peaks overlapped, and the dodecamer was eluted as a separate peak (Figure 5.28a). As the temperature increased, the peaks merged more. At 25°C, there was no dodecamer peak, and a large tetramer peak was merged with a small octamer peak (Figure 5.28d). The dynamic column profiles for 4°C and 25°C are plotted in Figures 5.29 and 5.30, respectively. For both cases, the injected sample had an equilibrium distribution of 22% tetramers, 60% octamers, and 18% dodecamers.

Figures 5.29a and 5.30a show that the distributions of the aggregates in the column at $t = 0.1$ min were similar to the equilibrium distribution. At 4°C, the reaction rates were relatively slow. As seen in Figure 5.29, the three aggregates tended to elute separately as individual species. As the dodecamer moved ahead of the other two aggregates, it tended to dissociate to tetramer and octamer, whereas the association reaction was inhibited by the separation of tetramer and octamer. In addition, the separation of the tetramer and octamer peaks also helped the dissociation of the octamers to form tetramers. Thus, the tetramer peak kept growing, whereas the octamer and the dodecamer peaks continued to shrink. The changes of the peak areas were relatively slow because of the slow reaction rates. The three aggregates were gradually resolved in the column since they had different affinities with the sorbent.

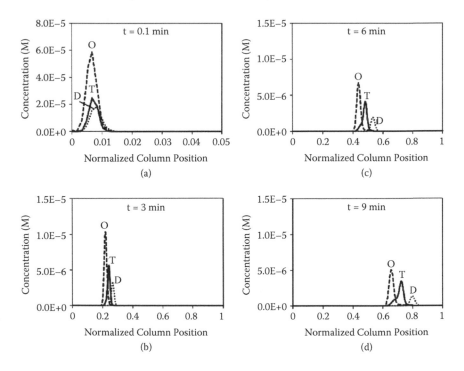

FIGURE 5.29 Dynamic column profiles of the chromatogram in Figure 5.28a.

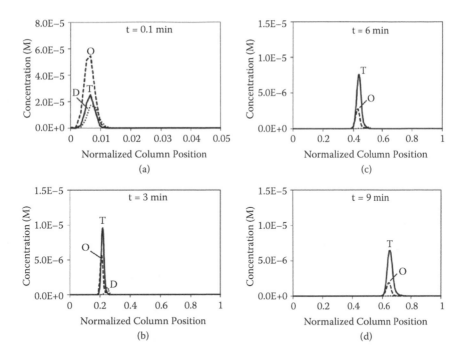

FIGURE 5.30 Dynamic column profiles of the chromatogram in Figure 5.28d.

At 25°C, the reaction rates were relatively high (all N_k values were greater than 1), and the different aggregates tended to merge and migrate at the same speed (see Figure 5.30). The two aggregation reactions (see Figure 5.1) reached equilibrium rapidly. The concentrations of the three species were related by the two equilibrium constants, $C_O = K_1 C_T^2$ and $C_D = K_2 C_T C_O = K_1 K_2 C_T^3$. Since the tetramer peak was diluted in the column, C_T decreased, leading to a faster decrease of C_O and C_D by dissociation. Hence, the two reactions occurred from right to left. During the migration, the tetramer peak kept growing, whereas the octamer and the dodecamer peaks shrunk rapidly. As shown in Figures 5.30c and 5.30d, after 6 min, the dodecamer peak was too small to be identified. Eventually, a large tetramer peak eluted together with a small octamer peak, and the distribution reached reaction equilibrium.

5.8.3 INHIBITION OF AGGREGATION

In many cases, aggregation of proteins or polymers in chromatography is undesirable for both analytical and production applications. One can reduce aggregation by reducing the concentration or by controlling the N_{k+} and N_{k-} values. In elution chromatography, peaks are diluted during migration in the column, resulting in a shift of the equilibrium toward the tetramer. In Figures 5.28a–5.28d, the tetramer conversion is reduced significantly as the temperature increases from 4°C to 25°C. Since the N_{k+} and N_{k-} values for both reactions at 25°C are greater than 1, the peaks merge, and reaction equilibrium is reached rapidly.

To increase N_{k+} and N_{k-}, one can either increase the temperature, to increase k_+ and k_- (Figures 5.28a–5.28d), or increase the HPLC residence time (L/u_0). Figure 5.28e shows that if the residence time is increased by 10-fold by decreasing the flow rate, the peaks of different aggregates also merge. The distribution is close to that in Figure 5.28d since Figure 5.28e has similar values of N_{k+} and N_{k-} for both reactions.

Figure 5.28f shows that a 10-fold decrease of the injection volume in Figure 5.28e lowers the tetramer peak concentration and thus reduces the aggregation significantly. The conversion of the tetramer is reduced to less than 0.04. Besides reducing the injection volume, additional peak dilution can also be obtained by decreasing the feed concentration or by increasing the particle size. The dilution methods, however, may cause detection problems or lower accuracy in quantitative analysis.

5.8.4 IMPROVEMENT OF RESOLUTION

If the target is to separate and analyze the different aggregates rather than inhibit the aggregation reactions, one should choose conditions that allow the presence of sufficient amounts of each aggregate and increase the peak resolution. To have considerable amounts of the larger aggregates, N_{k+} and N_{k-} have to be sufficiently low. Since the temperature cannot decrease much below 4°C, an alternative way to reduce N_{k+} and N_{k-} is decreasing the residence time (L/u_0). Figure 5.31a is a simulated chromatogram at 4°C in which the flow rate is increased 10-fold, and N_{k+} and N_{k-} are thus decreased by 10-fold compared to Figure 5.28a. The amounts of octamers and dodecamers in the effluent are much larger than those in Figure 5.28a, and the distribution of the aggregates approaches the equilibrium distribution in the feed, as shown in Figures 5.29a and 5.30a. The overlap of the tetramer with the octamer peak is reduced in Figure 5.31a, compared to Figure 5.28a, as N_{k-} is reduced from 0.211 to 0.0211. The peaks are spread more than those in Figure 5.28a since a higher velocity results in higher intraparticle diffusion resistances.

To improve the resolution of the different aggregates and reduce wave spreading, one can use smaller particles. Figures 5.31b–5.31d show how peak shapes vary with particle size when N_{k+} and N_{k-} are kept the same as those in Figure 5.31a. Apparently, when particle size decreases, N_D, Pe_b, and N_f decrease, and the resolution increases. When the particle radius is as low as 1.25 μm ($N_D \sim 10^3$, $Pe_b \sim 10^4$, $N_f \sim 10^5$), the three aggregates can be well resolved.

5.8.5 ESTIMATION OF THE REACTION STOICHIOMETRY
AND THE EQUILIBRIUM CONSTANTS

When the reaction rates are slower than the convection rate, the distribution of the different aggregates is similar to the distribution at equilibrium. Then, if the peaks are well resolved, one can estimate the aggregation numbers or the reaction stoichiometry from the peak areas (A_D, A_O, and A_T) at different feed concentrations. As the concentration increases, the relative peak areas of the larger aggregates (O and D) increase (Figure 5.32). If $\ln (A_D)$ or $\ln (A_O)$ is plotted against $\ln (A_T)$ at different feed concentrations, the slope is equal to the aggregation number of D or O relative to T [3].

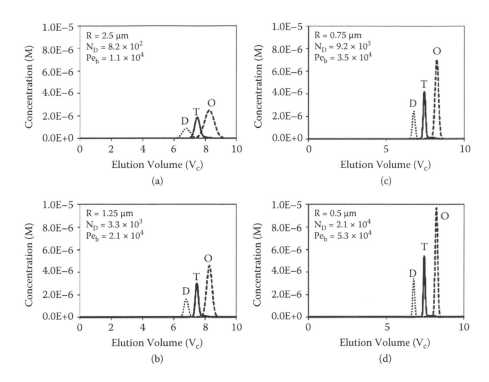

FIGURE 5.31 Effect of mass transfer resistances on resolution for Case V. The flow rate is 10 mL/min. The particle size is given in each figure. The reaction dimensionless groups are $N_{k+1} = 0.137$; $N_{k-1} = 0.021$; $N_{k+2} = 0.026$; $N_{k-2} = 0.029$; $N_f = 3.7 \times 10^4$ (a); 1.2×10^5 (b); 2.7×10^5 (c); 5.4×10^5 (d). Other parameters are the same as in Figure 5.28a. The distributions of tetramers, octamers, and dodecamers are 24%, 59%, and 17%, respectively.

Once the reaction stoichiometry is established, the equilibrium constants can be obtained from the peak ratios. If the extinction coefficients are proportional to the aggregation number, then A_T is proportional to C_T, A_O is proportional to $2C_O$, and A_D is proportional to $3C_D$, where C_T, C_O, and C_D are the equilibrium molar concentrations of tetramers, octamers, and dodecamers, respectively. Since $C_T + 2C_O + 3C_D = C_T + 2K_1C_T^2 + 3K_1K_2C_T^3 = C_{f,T}$, one can estimate K_1 and K_2 from the following equations:

$$C_T = C_{f,T} \left(\frac{A_T}{A_T + A_O + A_D} \right) \tag{5.27a}$$

$$K_1 = C_{f,T} \left(\frac{A_O}{A_T + A_O + A_D} \right) \frac{1}{2C_T^2} \tag{5.27b}$$

$$K_2 = C_{f,T} \left(\frac{A_D}{A_T + A_O + A_D} \right) \frac{1}{3K_1C_T^3} \tag{5.27c}$$

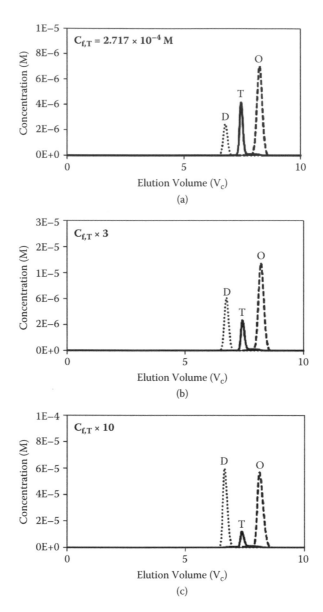

FIGURE 5.32 Effect of feed concentration on the distribution of different aggregates for Case V. $F = 10$ mL/min and $R = 0.75$ μm. The reaction dimensionless groups are $N_{k+,1} = 0.137$ (a); 0.412 (b); 1.373 (c); $N_{k+,2} = 0.026$ (a); 0.077 (b); 0.257 (c); $N_{k-,1} = 0.0211$; $N_{k-,2} = 0.0287$. Other parameters and dimensionless groups are the same as in Figure 5.31c.

5.8.6 SUMMARY

A system with two reversible aggregation reactions in the mobile phase was modeled and simulated. The simulated chromatograms agreed with the experimental data reported by Grinberg et al. [3]. The dynamic column profiles show that the distribution of the various aggregates can change during migration in a column. At the low temperature of 4°C, the reaction rates are relatively slow, the aggregates elute separately as individual species, and the changes of the aggregate distribution are relatively small. At the higher temperature of 25°C, the reaction rates are high, and the various aggregates can reach reaction equilibrium rapidly and migrate at the same speed. Since the peaks are spread because of mass transfer effects, octamers and dodecamers tend to dissociate to form tetramers. Hence, a large tetramer peak coeluts with a small octamer peak. During migration in the column, the dodecamers dissociate completely.

If the aggregation reactions are undesirable, one may increase the relative reaction rates by increasing the temperature or by increasing the residence time to lower the conversion of the smaller aggregates. One can also reduce the peak concentration by reducing the feed concentration or the injection volume or by increasing the particle size. When $N_{k-} > 2$ and $L_f \sim 10^{-4}$, the conversion of the tetramers is found to be less than 0.05. To resolve and analyze the various aggregates and to ensure that all the aggregates are present in the effluent, the relative reaction rates need to be kept small by decreasing the temperature or the residence time. To improve resolution, the particle size needs to be reduced. If $N_{k+} < 0.1$, $N_{k-} < 0.1$, $N_D > 10^3$, $Pe_b > 10^4$, and $N_f > 10^4$, the peaks of the aggregates for a small pulse ($L_f \sim 10^{-3}$) are well resolved. Under these conditions, one can estimate the reaction stoichiometry and the equilibrium constants from the relative peak areas at different feed concentrations.

5.9 CASE VI: REVERSIBLE COMPLEXATION REACTIONS IN THE MOBILE PHASE FOR LIGAND-ASSISTED SEPARATIONS

5.9.1 SYSTEM CONSIDERED AND MODEL TESTING

The enantiomers of many drugs or chemicals have different physiological effects on living organisms. Ma and coworkers [12] used highly sulfated β-cyclodextrin (S-β-CD) as a ligand in the mobile phase to separate two enantiomers of chiral amine hydrochloride salt on C8 and C18 reversed-phase columns. Before the enantiomers were injected, the column was preequilibrated with the mobile phase containing the ligand. The ligand formed complexes with the enantiomers with different stability constants, and the complexed solutes adsorbed with different affinities. The reaction and adsorption phenomena in the column are summarized in Figure 5.1, where L refers to the ligand S-β-CD, R the R-enantiomer, and S the S-enantiomer. LR and LS are the complexed forms of R and S. The forward and the reversed complexation rate constants are k_+ and k_- respectively. The ratio k_+/k_- is equal to the stability constant K_C. The adsorption equilibrium constant of a species is K_A.

The retention factor k' was found to be proportional to the overall distribution coefficient, which could be determined from the complexation and the adsorption equilibrium constants [12]. The final equations are as follows:

$$k'_R = \frac{1}{\varepsilon_t}\left(\frac{K_{A,R} + K_{C,R}K_{A,LR}\ [L]}{1 + K_{C,R}\ [L]}\right) \tag{5.28a}$$

$$k'_S = \frac{1}{\varepsilon_t}\left(\frac{K_{A,S} + K_{C,S}K_{A,LS}\ [L]}{1 + K_{C,S}\ [L]}\right) \tag{5.28b}$$

where $\varepsilon_t = \varepsilon_b + (1 - \varepsilon_b)\varepsilon_p$ is the total void fraction in the column, and $[L]$ is the ligand concentration, which is constant in a given run; $K_{A,R}$ and $K_{A,S}$ are the adsorption equilibrium constants of R and S, respectively; $K_{A,LR}$ and $K_{A,LS}$ are the adsorption equilibrium constants of LR and LS, respectively; and $K_{C,R}$ and $K_{C,S}$ are the stability constants of the ligand for R and S, respectively. Equation (5.28) applies to HPLC systems with linear adsorption isotherms when the complexation and dissociation reaction are at equilibrium.

The retention factors of R and S as a function of the ligand concentration are shown in Figure 5.4a of Ma et al. [12]. In the absence of the ligand, when $[L] = 0$, the retention factors of R and S are the same, indicating that $K_{A,R} = K_{A,S}$, and that the sorbent has no selectivity for R or S. $K_{A,R}$ and $K_{A,S}$ can be calculated from the retention factors k'_R and k'_S. A least square fitting is then used to find the values of $K_{C,R}$, $K_{C,S}$, $K_{A,LR}$, and $K_{A,LS}$ from the data of k'_R versus $[L]$ and k'_S versus $[L]$ (see Table 5.13). These values were used in the simulations discussed here. Other parameters in the simulations are listed in Table 5.14. The simulated results for the separation of R and S on a C8 column are shown in Figure 5.33. The peaks have the same widths and retention times as the experimental data, indicating that the model and the parameters are plausible.

5.9.2   IMPROVEMENT OF SELECTIVITY

The overall selectivity α of S over R in the presence of a mobile phase additive can be determined from Equation (5.28).

$$\alpha = \frac{k'_S}{k'_R} = \left(\frac{K_{A,S} + K_{C,S}K_{A,LS}\ [L]}{K_{A,R} + K_{C,R}K_{A,LR}\ [L]}\right)\left(\frac{1 + K_{C,R}\ [L]}{1 + K_{C,S}\ [L]}\right) \tag{5.29}$$

TABLE 5.13

Adsorption and Complexation Equilibrium Constants for Case VI with Adsorption of Complexed Solutes

Component	K_A	K_C
R	4.418	—
S	4.418	—
LR	0.660	86.9
LS	0.764	95.9

TABLE 5.14
Simulation Parameters for Case VI

System Parameters

L (cm)	ID (cm)	R (μm)	ε_b	ε_p
15	0.46	1.35	0.35	0.25

Operating Conditions

F (mL/min)	$C_{f,R}$ and $C_{f,S}$ (mM)	t_L (min)
1.5	1	0.01

Reaction Parameters

Reaction	k_+ (mM^{-1} min^{-1})	k_- (min^{-1})
$L + R \leftrightarrow LR$	1738	20
$L + S \leftrightarrow LS$	1918	20

Isotherm Parameters (Langmuir)

Component	a	b (mM^{-1})
R	4.418	0
S	4.418	0
L	4.418	0
LR	0.66	0
LS	0.764	0

Mass Transfer Parameters

Component	D_b (cm^2/min)	D_p (cm^2/min)	k_f (cm/min)	E_b (cm^2/min)
All Species	0.001	0.0005	From [37]	From [36]

Numerical Parameters

Axial Elements	Step Size (L/u_0)	Collocation Points Axial	Collocation Points Particle	Tolerance Absolute (mM)	Tolerance Relative
100	0.01	4	2	10^{-4}	10^{-4}

The selectivity depends on the concentration of the ligand $[L]$, the stability constants of the ligand for the solutes ($K_{C,R}$ and $K_{C,S}$), and the adsorption equilibrium constants of the complexed solutes and the free solutes ($K_{A,LR}$, $K_{A,LS}$, $K_{A,R}$, and $K_{A,S}$).

If no ligand is present, or $[L] = 0$, α is equal to sorbent selectivity for the free solutes.

$$\alpha = \frac{K_{A,S}}{K_{A,R}} \tag{5.30}$$

If some ligand is present, $[L] > 0$, and the complexed solutes do not adsorb, $K_{A,LR} = K_{A,LS} = 0$, the selectivity α in Equation (5.29) is reduced to

$$\alpha = \left(\frac{K_{A,S}}{K_{A,R}} \right) \left(\frac{1 + K_{C,R}\,[L]}{1 + K_{C,S}\,[L]} \right) \tag{5.31}$$

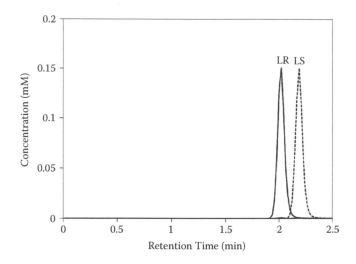

FIGURE 5.33 Simulated chromatogram for Case VI. *LR* is the complexed R-enantiomer, and *LS* is the complexed S-enantiomer. The concentrations of free *R* and *S* are negligible. The parameters used in the simulation are listed in Table 5.14. The dimensionless groups are $N_{k+,R} = 2.0 \times 10^3$, $N_{k+,S} = 2.2 \times 10^3$, $N_{k-,R} = N_{k-,S} = 11.6$, $N_D = 7.4 \times 10^3$, $Pe_b = 3.2 \times 10^4$, $N_f = 3.7 \times 10^5$, $L_f = 0.0015$.

If the products $K_{C,R}[L]$ and $K_{C,S}[L]$ are much larger than 1, then the overall selectivity is equal to the ratio of the sorbent selectivity to the ligand selectivity.

$$\alpha \cong \left(\frac{K_{A,S}}{K_{A,R}} \right) \left(\frac{K_{C,R}}{K_{C,S}} \right) \qquad (5.32)$$

A high overall selectivity can be obtained if the sorbent and the ligand have opposite relative affinities for the enantiomers. If the sorbent has no selectivity for the free solutes, $K_{A,S}/K_{A,R} = 1$, the overall selectivity is determined by the ligand selectivity $K_{C,S}/K_{C,R}$ and the ligand concentration $[L]$.

$$\alpha = \frac{1 + K_{C,R}\ [L]}{1 + K_{C,S}\ [L]} \qquad (5.33)$$

For large values of $[L]$, the overall selectivity α approaches the value of $K_{C,R}/K_{C,S}$, which is the reciprocal of the ligand selectivity for S.

If $[L] > 0$ and the complexed solutes can adsorb ($K_{A,LR}$, $K_{A,LS} > 0$), the overall selectivity is given by Equation (5.29). Generally, α is a complex function of $[L]$, $K_{A,S}/K_{A,R}$, $K_{C,S}/K_{C,R}$, and $K_{A,LS}/K_{A,LR}$.

If $[L]$ is small, then the overall selectivity α approaches the sorbent selectivity for the free solutes.

$$\alpha \approx \frac{K_{A,S}}{K_{A,R}} \qquad (5.34)$$

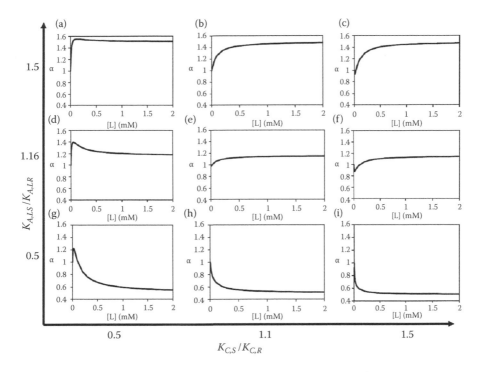

FIGURE 5.34 Relation of selectivity to ligand concentration for Case VI at various values of $K_{C,S}/K_{C,R}$, and $K_{A,LS}/K_{A,LR}$ (Equation 5.29). The ratio $K_{A,S}/K_{A,R}$ is assumed to be equal to 1.

If [L] is large, then α reduces to

$$\alpha \approx \left(\frac{K_{C,S}K_{A,LS} \; [L]}{K_{C,R}K_{A,LR} \; [L]} \right) \left(\frac{K_{C,R} \; [L]}{K_{C,S} \; [L]} = \frac{K_{A,LS}}{K_{A,LR}} \right) \tag{5.35}$$

Here, α is only a function of the sorbent selectivity for the complexed solutes, $K_{A,LS}/K_{A,LR}$. Figure 5.34 shows α versus [L] at different values of $K_{C,S}/K_{C,R}$ and $K_{A,LS}/K_{A,LR}$ when the sorbent has no selectivity for the free solutes or if $K_{A,S}/K_{A,R}= 1$. The selectivities at [L] $= 0$ mM and [L] $= 2$ mM are consistent with the predictions of Equation (5.34) and Equation (5.35). Graph e with $K_{C,S}/K_{C,R} = 1.10$ and $K_{A,LS}/K_{A,LR}= 1.16$ was based on the data of Ma et al. [12].

Since adsorption of the complexed solutes can affect significantly the overall selectivity [see Equation (5.29)], it is important to determine whether the complexed solutes can adsorb. Equation (5.28) can be rearranged to obtain

$$\frac{1}{k'_R} = \frac{\varepsilon_t(1+ K_{C,R} \; [L])}{K_{A,R} + K_{C,R}K_{A,LR} \; [L]} \tag{5.36a}$$

$$\frac{1}{k'_S} = \frac{\varepsilon_t(1+ K_{C,S} \; [L])}{K_{A,S} + K_{C,S}K_{A,LS} \; [L]} \tag{5.36b}$$

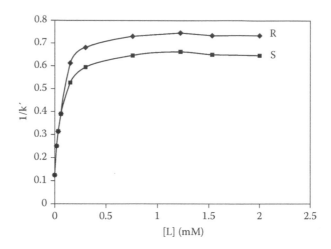

FIGURE 5.35 The relation between $1/k'$ and ligand concentration for Case VI. The data points were obtained from Ma et al. [12].

If the complexed solutes do not adsorb, then $K_{A,LS} = K_{A,LR} = 0$; a plot of $1/k'$ versus $[L]$ is linear. Otherwise, the plot is nonlinear. The data of Ma et al. indicate that the complexed solutes adsorb (Figure 5.35).

5.9.3 IMPROVEMENT OF RESOLUTION IF THE COMPLEXED SOLUTES ADSORB

If the complexed solutes can adsorb and the ligand concentration is sufficiently high, the overall selectivity is determined by the sorbent selectivity for the complexed solutes $K_{A,LS}/K_{A,LR}$ (see Equation 5.35). If the complexed solutes have different affinities with the sorbent, the selectivity is high. Nonetheless, a good selectivity is insufficient for obtaining a high resolution because of coupling of the chromatographic process with reactions. This section focuses on how reaction and mass transfer rates affect the resolution for the system described in Figure 5.33.

As the relative reaction rates increase, the resolution increases (Figure 5.36). When the value of N_{k+} is small (~10; Figure 5.36a), the complexation reaction rate is slow, and it takes a long time for the reaction to reach equilibrium. The solutes are mostly in the free forms during migration. Since the sorbent has a high affinity but no selectivity for the free solutes, the two peaks show significant overlap and tailing. When N_{k+} is high (>10^3; Figures 5.36c and 5.36d), the reaction rate is fast, and the complexed solutes are formed rapidly and eluted with a high resolution. The resolution increases with N_{k+} and approaches a constant when $N_{k+,R}$ is greater than 10^3 (see Figure 5.37a).

This value of N_{k+} is much higher than those required for complete reactions for Cases I and III ($N_{k+} > 5$) because the solutes can be well separated only if they are fully complexed with the ligand soon after they enter the column. Since the sorbent has a small selectivity (1.16) for the complexed solutes but no selectivity for the free solutes, such fast complexation allows a large portion of the column

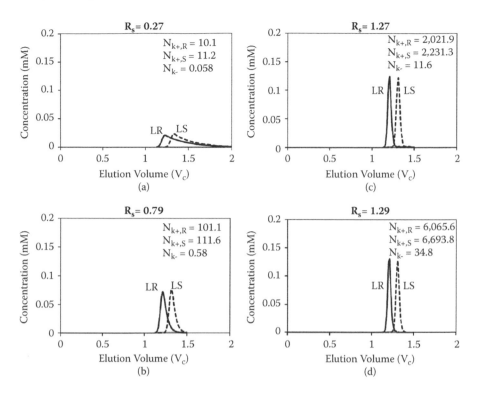

FIGURE 5.36 Effect of reaction rates on resolution for Case VI. N_{k+} and N_{k-} are varied by changing k_+ and k_- while keeping the ratio k_+/k_- constant. The values of N_{k-} for the two reactions are the same. Other parameters and dimensionless groups are given in Figure 5.33.

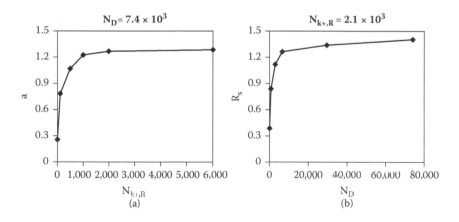

FIGURE 5.37 Effects of N_{k+} (a) and N_D (b) on resolution for Case VI. (a) N_D, Pe_b, and N_f are the same as in Figure 5.33, whereas N_{k-} changes proportionally with N_{k+} to keep K_C the same as in Figure 5.33. (b) N_{k+} and N_{k-} are the same as in Figure 5.33, whereas N_D is increased by reducing the particle size. Pe_b and N_f are also increased by reducing the particle size (not shown).

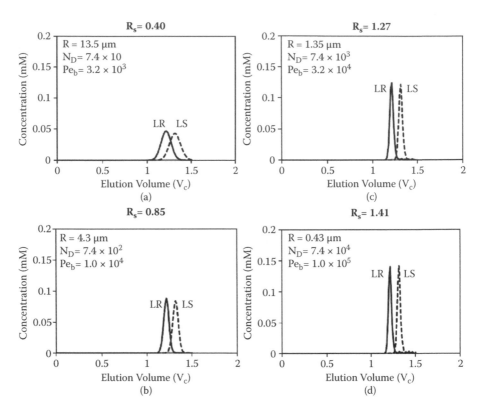

FIGURE 5.38 Effect of mass transfer resistances on resolution for Case VI. The particle size is given in each figure. $N_f = 3.7 \times 10^4$ (a); 5.5×10^5 (b); 4.7×10^5 (c); 2.5×10^6 (d). Other parameters and dimensionless group values are the same as in Figure 5.33.

to be used for resolving the complexed solutes. If the feed is mixed with the ligand and the complexation/dissociation reactions have already reached equilibrium before the sample injection, the entire column can be used to separate the complexed solutes.

The resolution increases with decreasing particle size (Figure 5.38). For $R = 13.5$ μm (Figure 5.38a), N_D, Pe_b, and N_f are small, and the resolution is low. Conversely, for $R = 1.35$ or 0.43 μm (Figures 5.38c and 5.38d), the resolution is high. The resolution increases rapidly with N_D at $N_D < 10^3$, where intraparticle diffusion controls peak spreading (Figure 5.37b). For N_D larger than 10^3, further gains in resolution are small. Axial dispersion ($Pe_b \sim 10^4$) and film diffusion ($N_f \sim 10^5$) also contribute to the peak spreading, but the overall mass transfer effect on the resolution is small.

Therefore, for the complexation reaction to reach equilibrium soon after sample injection, the N_{k+} value should be larger than 10^3. To maximize the resolution for the given selectivity, the following mass transfer parameter values are needed: $N_D > 10^3$, $Pe_b > 10^4$, and $N_f > 10^4$.

5.9.4   IMPROVEMENT OF RESOLUTION IF THE COMPLEXED
SOLUTES DO NOT ADSORB

For complexant-assisted separation of lanthanides in ion exchange columns, the complexed solutes do not adsorb since they have no net charges [13]. Then, $K_{A,LS} = K_{A,LR} = 0$, and the retention factor k' of Equation (5.28) can be reduced to

$$k'_R = \frac{1}{\varepsilon_t} \left(\frac{K_{A,R}}{1 + K_{C,R}\,[L]} \right) \tag{5.37a}$$

$$k'_S = \frac{1}{\varepsilon_t} \left(\frac{K_{A,S}}{1 + K_{C,S}\,[L]} \right) \tag{5.37b}$$

Here, R and S refer to two different lanthanide ions. If the ligand concentration is high, the overall selectivity α is equal to the ratio of the sorbent selectivity to the ligand selectivity by Equation (5.32). If the sorbent has no selectivity for the free solutes, then α is simply the reciprocal of the ligand selectivity ($K_{C,R}/K_{C,S}$).

In addition to high selectivity, high complexation reaction rates and low mass transfer resistances are important but insufficient to guarantee high resolution. According to Equation (5.37), if the stability constants K_C or the ligand concentration $[L]$ are large, the product $K_C[L]$ value can be much larger than K_A, and the retention factors for both R and S approach zero. Both solutes elute at the total void volume with poor resolution, as shown in an example in Figure 5.39.

In this example, the sorbent has no selectivity, and the ligand has a selectivity of 2 for S over R. The values $N_{k+} = 2.0 \times 10^3$ for R and 4.0×10^3 for S, and $N_D = 7.4 \times 10^3$. Other parameters are the same as those in Table 5.14. The column is preequilibrated with the mobile phase containing the ligand, which does not adsorb. When $K_C[L]$ is much larger than $K_{A,R}$ and $K_{A,S}$ (Figure 5.39a), the values k'_R and k'_S are less than 0.05 (Table 5.15). The two solutes are eluted close to the total void volume with poor resolution ($R_S = 0.27$). No peak is seen for the free solutes since all the solutes are complexed strongly with the ligand.

As $K_C[L]$ decreases, k'_R and k'_S increase, and the resolution R_S increases to 0.8 (Figure 5.39b). When $K_C[L]$ is comparable to the values of $K_{A,R}$ and $K_{A,S}$, the two solutes are well resolved ($R_S = 1.63$ in Figure 5.39c, and $R_S = 2.13$ for Figure 5.39d). Since both N_{k+} and N_{k-} are larger than 10 for R and S, the peaks of S and LS are merged into a single peak. Similarly, the peaks for R and LR are also merged. When $K_C[L]$ is much smaller than $K_{A,R}$ and $K_{A,S}$, the complexation is weak, resulting in longer retention times and broader peaks (Figure 5.39e). When $K_C[L]$ is much smaller than 1 as in Figure 5.39f, the complexation is negligible, and there is little overall selectivity, $\alpha(R/S) = 1.05$ (Table 5.15). Thus, the two solutes coelute in the free forms with poor resolution ($R_S = 0.63$). To achieve high resolution with minimum peak dilution, a ligand should have good selectivity (~2 or higher) and an optimal $K_C[L]$ value, which is of the same order of magnitude as K_A (Table 5.15).

In addition to the value of $K_C[L]$, which is equal to N_{k+}/N_{k-}, the values of N_{k+} and N_{k-} can also affect peak retention, spreading, and resolution. In Figure 5.39, the N_{k+} value is fixed at 2×10^3 for R and 4×10^3 for S, and the N_{k-} values are all larger than 10. Under these conditions, the complexation and dissociation reactions

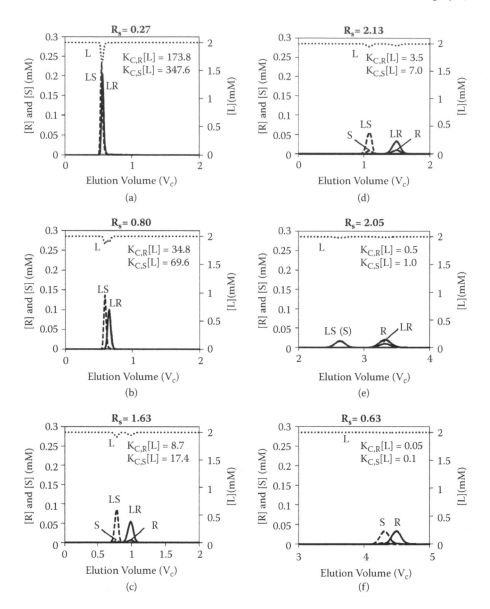

FIGURE 5.39 Effect of $K_C[L]$ on resolution if the complexed solutes do not adsorb ($K_{A,LR} = K_{A,LS} = 0$). $K_C[L]$ is varied by changing k_-, whereas k_+ and $[L]$ are kept constant. The sorbent has no selectivity, $K_{A,R} = K_{A,S} = 4.418$. The ligand has a selectivity of 2 for S over R ($K_{C,S}/K_{C,R} = 2$). $N_{k+,R} = 2.0 \times 10^3$, $N_{k+,S} = 4.0 \times 10^3$. $N_{k-,R} = N_{k-,S} = 11.6$ (a); 58 (b); 232 (c); 580 (d); 4060 (e); 40,600 (f). Other parameters and dimensionless group values are the same as in Figure 5.33.

TABLE 5.15
Effects of $K_C[L]$ on Retention Factor, Overall Selectivity, and Peak Resolution for Case VI without Adsorption of Complexed Solutes

$K_{A,S} = K_{A,R} = 4.418$, $K_{A,LS} = K_{A,LR} = 0$, $K_{C,S}/K_{C,R} = 2$, $N_{k+} \sim 10^3$, $N_{k-} > 10$.

Figure 5.39	$K_{C,R}[L]$	$K_{C,S}[L]$	k_R'	k_S'	α (R/S)	R_S
(a)	173.8	347.6	0.0493	0.0247	2.00	0.27
(b)	34.8	69.6	0.241	0.122	1.98	0.80
(c)	8.7	17.4	0.889	0.469	1.90	1.63
(d)	3.5	7.0	1.92	1.08	1.78	2.13
(e)	0.5	1.0	5.75	4.31	1.33	2.05
(f)	0.05	0.1	8.21	7.84	1.05	0.63

are fast, so that the retention times can be predicted from the retention factors of Equation (5.37), which assumes that the reactions are at equilibrium. Figure 5.40 shows the peak behaviors at different N_{k+} and N_{k-} values. When the reactions are too slow to reach equilibrium ($N_{k-} \sim 1$; Figure 5.40a), the complexed solutes tend to elute at the total void volume, whereas the free solutes tend to elute much later

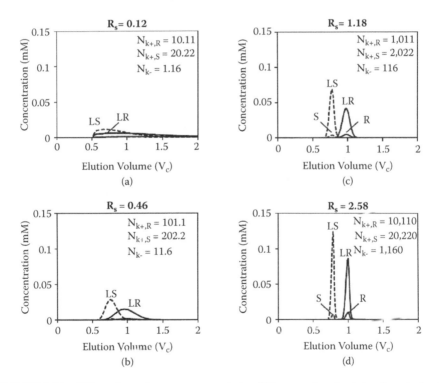

FIGURE 5.40 Effect of reaction rates on resolution if the complexed solutes do not adsorb ($K_{A,LR} = K_{A,LS} = 0$). $K_{A,R} = K_{A,S} = 4.418$, $K_{C,R}[L] = 8.7$, and $K_{C,R}[L] = 17.4$. N_{k+} and N_{k-} are varied by changing k_+ and k_-, respectively. Other parameters and dimensionless group values are the same as in Figure 5.33.

(about 4.5 V_C). As a result, the peaks of LS and LR are broadened significantly, and the retention times deviate from those given by Equation (5.37). As N_{k-} increases to 10 (Figure 5.40b), the peak retention times are consistent with the prediction by Equation (5.37), but the peaks are spread because of the finite reaction rates. When N_{k-} increases to larger than 10^2 (Figures 5.40c and 5.40d), the spreading due to reactions is reduced, and the resolution is high. In this system, the peak spreading due to the reactions are evident because N_{k-} values are smaller than those of N_D and Pe_b.

5.9.5 SUMMARY

The separation of chiral amine hydrochloride salts using S-β-CD as a mobile phase additive was analyzed and simulated. The complexed solutes can adsorb in this system. The predicted peak retention times and peak widths were consistent with the data [12]. The adsorption of the complexed solutes has a large effect on the overall selectivity. At a high ligand concentration, the selectivity approaches the sorbent selectivity for the complexed solutes, $K_{A,LS}/K_{A,LR}$. At a low ligand concentration, the selectivity is determined by the sorbent selectivity for the free solutes $K_{A,S}/K_{A,R}$, the ligand selectivity $K_{C,S}/K_{C,R}$, and the sorbent selectivity for the complexed solutes $K_{A,LS}/K_{A,LR}$.

If the complexed solutes do not adsorb, as for complexant-assisted separation of lanthanides in an ion exchange column, the selectivity increases with the ligand concentration and eventually approaches the ratio of the sorbent selectivity $K_{A,S}/K_{A,R}$ to the ligand selectivity $K_{C,S}/K_{C,R}$. If the adsorbent has a lower value of the adsorption equilibrium constant K_A for the solute preferred by the ligand, the adsorbent and the ligand work synergistically to result in better separation. If $1/k'$ is a linear function of $[L]$, the complexed solutes do not adsorb.

To reduce HPLC peak spreading, which causes lower resolution, one should increase the relative reaction rate and reduce mass transfer resistances. The conditions for achieving a high resolution are $N_{k+} > 10^3$, $N_D > 10^3$, $Pe_b > 10^4$, and $N_f > 10^4$.

If the complexed solutes do not adsorb and the sorbent has no selectivity for the free solutes, the product $K_C[L]$ of the stability constant and the ligand concentration should be optimized for achieving good resolution. If $K_C[L]$ is too large ($>>K_A$), the retention factors are too small. Both solutes coelute rapidly at the total void volume with a poor resolution even when the selectivity is high and the peaks are sharp. If $K_C[L]$ is too small ($<<1$), the solute retention factors are large, and the overall selectivity approaches the sorbent selectivity, which is generally small. If $K_C[L]$ is of the same order of magnitude as the adsorption equilibrium constant K_A for the free solutes, the solute preferred by the ligand will migrate faster than the other, and the two solute peaks will be well resolved. To reduce significant peak spreading due to the slow reactions, the N_k values should be larger than those of N_D and Pe_b.

5.10 CONCLUSIONS

Various chromatographic systems with reactions in the mobile phase have been modeled using VERSE simulations and interpreted with dimensionless groups. The results provide overall guidelines for understanding, designing, and optimizing chromatographic processes with reactions. One can design experiments to determine whether reactions occur and the kinetics and equilibria of such reactions. For an irreversible first-order reaction, the rate constant can be obtained from the peak areas as a function of the residence time (Case I). For a reversible first-order reaction, the equilibrium constant can be estimated from the peak ratios (Case II). For large-scale production using chromatography reactors, a reversible decomposition reaction can overcome the equilibrium limit and approach complete conversion if the reaction rate is high compared to the convection rate. The loading volume should be sufficiently small to ensure the separation of the product bands (Case III). High-purity products can be obtained when the diffusion rate is large and the axial dispersion rate is small relative to the convection rate.

For reversibly aggregating systems, if the reactions reach equilibrium, one can separate the various aggregates in SEC [Case IV(A)] or adsorption chromatography (Case V) by decreasing the reaction rates relative to the convection rate. This can be achieved by decreasing the reaction temperature or the residence time. The reaction stoichiometry and the equilibrium constants can be estimated from the HPLC peak areas. In capture chromatography [Case IV(B)], one can merge multiple wave fronts and increase the dynamic binding capacity by increasing the reaction rate relative to the convection rate.

For ligand-assisted separations (Case VI), if the complexed solutes can adsorb, the overall selectivity approaches the sorbent selectivity for the complexed solutes when the ligand concentration is sufficiently high. If the complexed solutes do not adsorb, the overall selectivity at a high ligand concentration is equal to the ratio of the sorbent selectivity to the ligand selectivity. Thus, to have synergistic effects on separation, the adsorbent selectivity should be opposite that of the ligand selectivity. The product of the stability constant and the ligand concentration should be kept of the same order of magnitude as the adsorption equilibrium constant for the free solutes so that the solutes can be retained and well separated in the column. To reduce significant peak spreading due to the slow reactions, the relative reaction rates should be larger than the relative mass transfer rates.

Chromatography systems with reactions in the mobile phase have many important applications. Moreover, when undesirable reactions occur, they may cause misinterpretation of the chromatograms. Theoretical analyses based on the dimensionless groups are useful for understanding the complex dynamic phenomena in such systems. One can also use the key dimensionless groups for designing and developing more efficient reaction and separation processes. This review was intended to facilitate new applications of chromatography with reactions.

ACKNOWLEDGMENT

First author Lei Ling is supported by the U.S. Department of Energy under contract DE-AC02-06CH11357.

NOTATION

a	Linear Langmuir linear isotherm parameter
a^*	Linear Freundlich isotherm parameter, M^{-m}
A	Peak area
A_J	First-derivative discretization matrix involving Jacobi polynomials

A_L	First-derivative discretization matrix involving Legendre polynomials
B_J	Second-derivative discretization matrix involving Jacobi polynomials
B_L	Second-derivative discretization matrix involving Legendre polynomials
b	Nonlinear Langmuir isotherm parameter, M^{-1} or mM^{-1}
Bi	Biot number, Table 5.3
c_b	Dimensionless bulk phase concentration
c_p	Dimensionless pore phase concentration
$\overline{c_p}$	Dimensionless solid phase concentration
C	Concentration, M or mM
C_b	Bulk phase concentration, M or mM
C_{br}	Breakthrough concentration, M or mM
C_f	Maximum concentration in the feed, M or mM
C_{in}	Concentration at the inlet of CSTR extracolumn dead volume, M or mM
C_{out}	Concentration at the outlet of CSTR extracolumn dead volume, M or mM
C_p	Pore phase concentration, M or mM
$\overline{C_p}$	Solid phase concentration, $mol\ L^{-1}$ or $mmol\ L^{-1}$ packing volume
$\overline{C_T}$	Maximum sorbent capacity, $mol\ L^{-1}$ or $mmol\ L^{-1}$ packing volume
D_b	Brownian diffusivity, $cm^2\ min^{-1}$
D_{BC}	Dynamic binding capacity, Equation (5.26)
$D_{BC}{}^*$	Dimensionless dynamic binding capacity, Equation (5.26)
D_p	Intraparticle pore diffusivity, $cm^2\ min^{-1}$
E_b	Axial dispersion coefficient
F	Flow rate, $mL\ min^{-1}$
HETP	Theoretical plate height, cm
J	Defined by Equation (5.10a)
k_+	Forward reaction rate constant (units depend on reaction)
k_-	Reverse reaction rate constant (units depend on reaction)
k'	Retention factor
k_{app}	Apparent reaction rate constant, min^{-1}
k_f	Film mass transfer coefficient, $cm\ min^{-1}$
k_m	Reaction rate constant in the mobile phase, min^{-1}
k_s	Reaction rate constant in the solid phase, min^{-1}
K	Reaction equilibrium constant (units depend on reaction)
K_A	Adsorption equilibrium constants
K_C	Stability constants, mM^{-1}
Ke	Size exclusion factor
L	Column packing length, cm
$[L]$	Ligand concentration, mM
L_f	Loading factor, Equations (5.13) and (5.14)
m	Nonlinear Freundlich isotherm parameter
n_a	Number of interior axial collocation points
n_p	Number of interior particle collocation points
N_D	N_pP, Table 5.3
N_f	Ratio of film mass transfer rate to convection rate, Table 5.3
$N_{k\pm}$	Ratio of reaction rate to convection rate, Table 5.2
N_p	Ratio of intraparticle diffusion rate to convection rate, Table 5.3
P	Phase ratio, $(1-\varepsilon_b)/\varepsilon_b$
$1/Pe_b$	Ratio of axial dispersion rate to convection rate, Table 5.3
P_r	Productivity, $mol\ min^{-1}\ L^{-1}$
q_f	Solid phase concentration in equilibrium with C_f, $mol\ L^{-1}$ or $mmol\ L^{-1}$ packing volume
r	Particle position, cm

R	The radius of sorbent particle, cm
Re	Reynold's number, Equation (5.9b)
R_S	Peak resolution
Sc	Schmidt number, Equation (5.10b)
t_{cycle}	Cycle time, min
t_L	Loading time, min
t_m	Retention time in the mobile phase, min
t_R	Total retention time, min
t_s	Retention time in the solid phase, min
u_0	Interstitial velocity, u_s/ε_b, cm min^{-1}
u_s	Superficial velocity, cm min^{-1}
V_C	Column volume, mL
V_D	Extracolumn dead volume, mL
V_f	Feed volume, L
x	Dimensionless column position
X	Conversion
Y	Yield, %
Y_b	Generation term by reaction in bulk phase
Y_l	The net loss in the pore phase by adsorption onto the solid phase
Y_p	Generation term by reaction in pore phase
z	Column position, cm

GREEK LETTERS

α	Selectivity
ε	Extinction coefficient
ε_b	Interparticle void fraction
ε_p	Intraparticle void fraction (porosity)
ε_t	Total void fraction in the column, $= \varepsilon_b + (1 - \varepsilon_b)\varepsilon_p$
θ	Dimensionless time
μ	Solution viscosity, g cm^{-1} min^{-1}
ξ	Dimensionless particle position
ρ	Solution density, g mL^{-1}
σ	Coefficient of component in the reaction rate equation
τ	Residence time, L/u_0, min

SUBSCRIPTS

h	A specific collocation point within either an axial element or the particle
i, j	Component counter
k	Collocation point counter within the same element as h
m	Reaction counter

REFERENCES

1. Bolme, M. W. and Langer, S. H. (1983) The liquid chromatographic reactor for kinetic studies, *Journal of Physical Chemistry* 87, 3363–3366.
2. LoBrutto, R., Bereznitski, Y., Novak, T. J., DiMichele, L., Pan, L., Journet, M., Kowal, J., and Grinberg, N. (2003) Kinetic analysis and subambient temperature on-line on-column derivatization of an active aldehyde, *Journal of Chromatography A* 995, 67–78.

3. Grinberg, N., Blanco, R., Yarmush, D. M., and Karger, B. L. (1989) Protein aggregation in high-performance liquid chromatography: hydrophobic interaction chromatography of beta-lactoglobulin A, *Analytical Chemistry 61*, 514–520.

4. Wetherold, R., Wissler, E., and Bischoff, K. (1975) An experimental and computational study of the hydrolysis of methyl formate in a chromatographic reactor, *Advances in Chemistry Series 133*, 181–190.

5. Cho, B., Carr, R., and Aris, R. (1980) A continuous chromatographic reactor, *Chemical Engineering Science 35*, 74–81.

6. Cho, B., Carr, R., and Aris, R. (1980) A new continuous flow reactor for simultaneous reaction and separation, *Separation Science and Technology 15*, 37–41.

7. Schweich, D. and Villermaux, J. (1982) The preparative chromatographic reactor revisited, *The Chemical Engineering Journal 24*, 99–109.

8. Hashimoto, K., Adachi, S., Noujima, H., and Ueda Y. (1983) A new process combining adsorption and enzyme reaction for producing higher-fructose syrup, *Biotechnology and Bioengineering 25*, 2371–2393.

9. Owens, P. K., Fell, A. F., Coleman, M. W., and Berridge, J. C. (1997) Chiral recognition in liquid chromatography utilising chargeable cyclodextrins for resolution of doxazosin enantiomers, *Chirality 9*, 184–190.

10. León, A. G., Olives, A. I., Martín, M. A., and Castillo, B. (2007) The role of β-cyclodextrin and hydroxypropyl β-cyclodextrin in the secondary chemical equilibria associated to the separation of β-carbolines by HPLC, *Journal of Inclusion Phenomena and Macrocyclic Chemistry 57*, 577–583.

11. León, A. G., Olives, A. I., Del Castillo, B., and Martín, M. A. (2008) Influence of the presence of methyl cyclodextrins in high-performance liquid chromatography mobile phases on the separation of beta-carboline alkaloids, *Journal of Chromatography A 1192*, 254–258.

12. Ma, S., Shen, S., Haddad, N., Tang, W., Wang, J., Lee, H., Yee, N., Senanayake, C., and Grinberg, N. (2009) Chromatographic and spectroscopic studies on the chiral recognition of sulfated beta-cyclodextrin as chiral mobile phase additive enantiomeric separation of a chiral amine, *Journal of Chromatography A 1216*, 1232–1240.

13. Powell, J. E. (1964) The separation of rare earths by ion exchange, in *Progress in the Science and Technology of the Rare Earths* (L. Eyring, Ed.), Macmillan, New York.

14. Jacobson, J., Melander, W., Vaisnys, G., and Horváth, C. (1984) Kinetic study on cis-trans proline isomerization by high-performance liquid chromatography, *Journal of Physical Chemistry 88*, 4536–4542.

15. Van Cott, K. E., Whitley, R. D., and Wang, N.-H. L. (1991) Effects of temperature and flow rate on frontal and elution chromatography of aggregating systems, *Separations Technology 1*, 142–152.

16. Whitley, R., Van Cott, K. E., Berninger, J. A., and Wang, N.-H. L. (1991) Effects of protein aggregation in isocratic nonlinear chromatography, *AIChE Journal 37*, 555–568.

17. Whitley, R., Zhang, X., and Wang, N.-H. L. (1994) Protein denaturation in nonlinear isocratic and gradient elution chromatography, *AIChE Journal 40*, 1067–1081.

18. Berninger, J. A., Whitley, R. D., Zhang, X., and Wang, N.-H. L. (1991) A versatile model for simulation of reaction and nonequilibrium dynamics in multicomponent fixed-bed adsorption processes, *Computers & Chemical Engineering 15*, 749–768.

19. Yu, Q. and Wang, N.-H. L. (1989) Computer simulations of the dynamics of multicomponent ion exchange and adsorption in fixed beds—gradient-directed moving finite element method, *Computers & Chemical Engineering 13*, 915–926.

20. Whitley, R. D., Van Cott, K. E., and Wang, N.-H. L. (1993) Analysis of nonequilibrium adsorption/desorption kinetics and implications for analytical and preparative chromatography, *Industrial & Engineering Chemistry Research 32*, 149–159.

21. Ma, Z., Whitley, R. D., and Wang, N.-H. L. (1996) Pore and surface diffusion in multicomponent adsorption and liquid chromatography systems, *AIChE Journal 42*, 1244–1262.
22. Koh, J.-H., Wang, N.-H. L., and Wankat, P. C. (1995) Ion exchange of phenylalanine in fluidized/expanded beds, *Industrial & Engineering Chemistry Research 34*, 2700–2711.
23. Koh, J.-H., Wankat, P. C. and Wang N.-H. L. (1998) Pore and surface diffusion and bulk-phase mass transfer in packed and fluidized beds, *Industrial & Engineering Chemistry Research 37*, 228–239.
24. Ernest, M. V., Whitley, R. D., Ma, Z., and Wang, N.-H. L. (1997) Effects of mass action equilibria on fixed-bed multicomponent ion-exchange dynamics, *Industrial & Engineering Chemistry Research 36*, 212–226.
25. Ernest, M. V., Bibler, J. P., Whitley, R. D., and Wang, N.-H. L. (1997) Development of a carousel ion-exchange process for removal of cesium-137 from alkaline nuclear waste, *Industrial & Engineering Chemistry Research 36*, 2775–2788.
26. Hritzko, B. J., Walker, D. D., and Wang, N.-H. L. (2000) Design of a carousel process for cesium removal using crystalline silicotitanate, *AIChE Journal 46*, 552–564.
27. Chin, C. Y. and Wang, N.-H. L. (2013) Simulated moving bed technology for biorefinery applications, in *Separation and Purification Technologies in Biorefineries* (S. Ramaswamy, H. Huang, and B. Ramarao, Eds.), Wiley, New York, DOI: 10.1002/9781118493441.
28. Yu, C.-M., Mun, S., and Wang, N.-H. L. (2006) Theoretical analysis of the effects of reversible dimerization in size exclusion chromatography, *Journal of Chromatography A 1132*, 99–108.
29. Tsui, H.-W., Hwang, M. Y., Ling, L., Franses, E. I., and Wang, N.-H. L. (2013) Retention models and interaction mechanisms of acetone and other carbonyl-containing molecules with amylose tris[(S)-α-methylbenzylcarbamate] sorbent, *Journal of Chromatography A 1279*, 36–48.
30. Yu, Q. and Wang, N.-H. L. (1986) Multicomponent interference phenomena in ion exchange columns, *Separation and Purification Methods 15*, 127–158.
31. Chung, P.-L., Bugayong, J. G., Chin, C. Y., and Wang, N.-H. L. (2010) A parallel pore and surface diffusion model for predicting the adsorption and elution profiles of lispro insulin and two impurities in gradient-elution reversed phase chromatography, *Journal of Chromatography A 1217*, 8103–8120.
32. Xie, Y., Farrenburg, C. A., Chin, C. Y., Mun, S., and Wang, N.-H. L. (2003) Design of SMB for a nonlinear amino acid system with mass-transfer effects, *AIChE Journal 49*, 2850–2863.
33. Xie, Y., Mun, S.-Y., and Wang, N.-H. L. (2003) Startup and shutdown strategies of simulated moving bed for insulin purification, *Industrial & Engineering Chemistry Research 42*, 1414–1425.
34. Xie, Y., Hritzko, B., Chin, C. Y., and Wang, N.-H. L. (2003) Separation of FTC-ester enantiomers using a simulated moving bed, *Industrial & Engineering Chemistry Research 42*, 4055–4067.
35. Lee, H.-J., Xie, Y., Koo, Y.-M., and Wang, N.-H. L. (2004) Separation of lactic acid from acetic acid using a four-zone SMB, *Biotechnology Progress 20*, 179–192.
36. Chung, S. F. and Wen, C. Y. (1968) Longitudinal dispersion of liquid flowing through fixed and fluidized beds, *AIChE Journal 14*, 857–866.
37. Wilson, E. J. and Geankoplis, C. J. (1966) Liquid mass transfer at very low Reynolds numbers in packed beds, *Industrial & Engineering Chemistry Fundamentals 5*, 9–14.
38. Mackie, J. S. and Meares, P. (1955) The diffusion of electrolytes in a cation-exchange resin membrane. *Proceedings of the Royal Society of London Series A 232*, 498–518.
39. Guiochon, G., Felinger, A., Shirazi, D. G., and Katti, A. M. (2006) *Fundamentals of Preparative and Nonlinear Chromatography* (2nd edition), Elsevier, San Diego, CA.
40. Wu, D. J., Xie, Y., Ma, Z., and Wang, N.-H. L. (1998) Design of simulated moving bed chromatography for amino acid separations, *Industrial and Engineering Chemistry Research 37*, 4023–4035.

41. Hritzko, B. J., Xie, Y., Wooley, R. J., and Wang, N.-H. L. (2002) Standing-wave design of tandem SMB for linear multicomponent systems, *AIChE Journal 48*, 2769–2787.
42. Lee, K., Chin, C. Y., and Xie, Y. (2005) Standing-wave design of a simulated moving bed under a pressure limit for enantioseparation of phenylpropanolamine, *Industrial and Engineering Chemistry Research 44*, 3249–3267.
43. Xie, Y., Chin, C. Y., Phelps, D. S. C., Lee, C.-H., Lee, K. B., Mun, S., and Wang, N.-H. L. (2005) A five-zone simulated moving bed for the isolation of six sugars from biomass hydrolyzate, *Industrial & Engineering Chemistry Research 44*, 9904–9920.
44. Mun, S., Xie, Y., and Wang, N.-H. L. (2005) Strategies to control batch integrity in size-exclusion simulated moving bed chromatography, *Industrial & Engineering Chemistry Research 44*, 3268–3283.
45. Chin, C. Y., Xie, Y., Alford, J., and Wang, N.-H. L. (2006) Analysis of zone and pump configurations in simulated moving bed purification of insulin, *AIChE Journal 52*, 2447–2460.
46. Mun, S., Wang, N.-H. L., Koo, Y.-M., and Yi, S. C. (2006) Pinched wave design of a four-zone simulated moving bed for linear adsorption systems with significant mass-transfer effects, *Industrial & Engineering Chemistry Research 45*, 7241–7250.
47. Villadsen, J., and Stewart, W. (1967) Solution of boundary-value problems by orthogonal collocation, *Chemical Engineering Science 22*, 1483–1501.
48. Carey, G. and Finlayson, B. (1975) Orthogonal collocation on finite elements, *Chemical Engineering Science 30*, 587–596.
49. Petzold L. R. (1982) *DASSL: A Differential/Algebraic System Solver*, Lawrence Livermore National Laboratory, Livermore, CA.
50. Chu, A. and Langer, S. (1985) Characterization of a chemically bonded stationary phase with kinetics in a liquid chromatographic reactor, *Analytical Chemistry 57*, 2197–2204.
51. Chu, A. H. T. and Langer, S. H. (1986) Measurement of reaction rate constants in the liquid chromatographic reactor: mass transfer effects, *Analytical Chemistry 58*, 1617–1625.
52. Jeng, C. Y. and Langer, S. H. (1992) Review reaction kinetics and kinetic processes in modern liquid chromatographic reactors, *Journal of Chromatography 589*, 1–30.
53. Melander, W., Jacobson, J., and Horváth, C. (1982) Effect of molecular structure and conformational change of proline-containing dipeptides in reversed-phase chromatography, *Journal of Chromatography 234*, 269–276.
54. Melander, W., Lin, H., Jacobson, J., and Horváth, C. (1984) Dynamic effect of secondary equilibria in reversed-phase chromatography, *Journal of Physical Chemistry 88*, 4527–4536.
55. Yu, C.-M., S. Mun, and Wang, N.-H. L. (2008) Phenomena of insulin peak fronting in size exclusion chromatography and strategies to reduce fronting, *Journal of Chromatography A 1192(1)*, 121–129.
56. Wong, J. W. (1990) Immobilized metal affinity chromatography, MS thesis, Purdue University, West Lafayette, IN.

Index

A

T - #0404 - 071024 - C4 - 234/156/12 - PB - 9780367378776 - Gloss Lamination